BRITISH OPENCAST COAL

A Photographic History 1942–1985

Old Pond
PUBLISHING

BRITISH OPENCAST COAL

A Photographic History
1942–1985

Keith Haddock

Old Pond
PUBLISHING

First published 2015

Copyright © Keith Haddock 2015

All rights reserved. No part of this publication may be reproduced, stored in a retrieval system, or transmitted, in any form or by any means, electronic, mechanical, photocopying, recording or otherwise, without prior permission of the copyright holder.

Published by
5M Publishing Ltd,
Benchmark House,
8 Smithy Wood Drive,
Sheffield, S35 1QN, UK
Tel: +44 (0) 1234 81 81 80
www.5mpublishing.com

A catalogue record for this book is available from the British Library

ISBN 978-1-910456-07-1

Book layout by Greg Sweetnam Design Solutions
Printed by Replika Press Pvt. Ltd
Photos from Keith Haddock's collection or by Keith Haddock (KH) unless otherwise stated.

BRITISH OPENCAST COAL

Contents

Author's biography: Keith Haddock — 7

Acknowledgements — 9

Introduction — 11

1 **The Early Years** — 13

2 **The 1950s** — 33

3 **The 1960s** — 59

4 **1970 to 1985** — 97

5 **Walking Draglines** — 125

6 **Hydraulic Excavators** — 135

7 **Significant Sites** — 151

Epilogue — 237

BRITISH OPENCAST COAL

Author's biography: Keith Haddock

Author Keith Haddock earned his qualifications as a Member of the Institution of Civil Engineers (MICE), Chartered Engineer (C.Eng.), Fellow of the Canadian Institute of Mining (FCIM) and registered professional engineer in Alberta (P.Eng.), but he prefers to be called an 'earthmover'. Over his long career in earthmoving and surface mining, he has become an expert authority on earthmoving equipment. Born in Sheffield, England, and graduating as a professional engineer there, he worked for major earthmoving contractors in the UK on highways, dams and opencast coal mining. In 1974, he moved to Canada and worked for Luscar Ltd., a large surface coal-mining company, where he stayed for 24 years, latterly as manager of engineering.

Keith has been responsible for several Alberta historical preservation projects at the Diplomat Mine Interpretive Centre, the Reynolds Alberta Museum and the Heavy Construction Heritage Society of Canada. In 1985, he co-founded the Historical Construction Equipment Association (HCEA) based in Bowling Green, Ohio, which now boasts about 5,000 members worldwide.

Since 1998 Keith has been a full-time author and freelance writer on heavy equipment. His articles have appeared in *Construction Equipment, Engineering News Record, Canadian Heavy Equipment Guide, Tracks & Treads* (Finning, Caterpillar), *Earthmovers* (UK), *Classic Plant & Machinery* (UK), *Cranes Today* (UK) and *Boggi* (Sweden). He is also known for his TV appearances on the History, Discovery and Learning Channels, on which he also acted as research consultant for *Monster Machines, Modern Marvels, Mega Excavators* and others. He is pleased to present this, his twelfth book, which takes him back to his childhood roots and the early years of his career.

Publications

Keith's articles and stories on construction and earthmoving equipment currently appear in the following magazines on a regular basis:
- *Tracks & Treads*, Finning house magazine (Canada)
- *Earthmovers* (UK)
- *Classic Plant & Machinery* (UK)
- *Equipment Echoes*, HCEA magazine (USA)

Keith's articles and stories have appeared periodically in the following magazines:
- *Canadian Heavy Equipment Guide* (Canada)
- *Boggi* (Sweden)
- *Engineering News Record* (USA)
- *Construction Equipment Magazine* (USA)
- *Coal News* (USA)
- *Cranes Today* (UK)

Books authored:
- *Caterpillar Modern Earthmoving Marvels*, jointly authored with Frank Raczon, MBI
- *Modern Earthmoving Machines at Work*, Iconografix
- *Earthmover Encyclopedia*, MBI
- *Bucyrus Heavy Equipment 1880–2008*, Iconografix
- *Bucyrus – Making the Earth Move for 125 Years*, MBI
- *Giant Earthmovers*, MBI
- *Classic Caterpillar Crawlers*, jointly authored with Eric C. Orlemann, MBI
- *Colossal Earthmovers*, MBI
- *Heavy Equipment,* jointly authored with Michael Alves, Hans Halberstadt and Sam Sargent, MBI
- *Extreme Mining Machines*, MBI
- *Marion Power Shovel 100 Years,* Marion

Assisted with the following books:
- *Coal in Canada*, CIM publication
- *Roads to Resources*, ARHCA publication
- *Coal Mining Equipment at Work,* (Michael Davis,) Iconografix
- *Marion Construction Machinery*, HCEA/Iconografix
- *Marion Mining & Dredging Machines*, HCEA/Iconografix
- *Erie Shovel Photo Archive*, HCEA/Iconografix
- *Opencast Coal Mining in Britain*, Peter Grimshaw, BCO
- *Amazing Story of Excavators*, Peter Grimshaw, KHL
- *Excavators*, Peter Grimshaw, Blandford

TV Shows

TV appearances, historical research and editing for the following documentaries:

History Channel:
- *Modern Marvels – Earthmovers* (1997),
- *Things that Move* (2006),
- *Tactical to Practical* (2004),
- *Modern Marvels – World's Biggest Machines* (2002),
- *Modern Marvels – Earthmovers II* (2005)

Discovery Channel:
- *Mega Excavators* (2002),
- *The Greatest Ever* (2005)

Learning Channel:
- *Monster Machines* (2000)

DVD Videos

Directed and authored *Massive Earthmoving Machines – Parts I & II*, two 70-minute videos on surface coal-mining in the USA and Germany. Published by Old Pond Publishing, Ipswich.

Directed and authored *Extreme Earthmovers at Work*, a 67-minute video on some of America's largest earthmoving machines including the Silver Spade stripping shovel. Filmed by Eric C. Orlemann, published by Old Pond Publishing, Ipswich.

BRITISH OPENCAST COAL

Acknowledgments

I wish to thank the many people who supported me in my younger years and encouraged me to pursue my career in civil engineering and earthmoving. I fondly remember these individuals, many of whom unfortunately are no longer with us, who allowed me to visit their opencast sites and took time to explain their operations in detail. There isn't room to mention everyone who helped me, but a few deserve special mention. In the early years of my career: Eric Grayson, Dennis Bateman and Jimmy Stringfellow of Northern Strip Mining Ltd.; John Foudy and Adam Umbrasko of G. Wimpey & Co. Ltd.; Terry Kennedy of Robert McGregor Ltd.; Ned Herrity of Murphy Brothers Ltd.; Ted Beer of Simms Sons & Cooke Ltd.; and Barry Gould and John Yeo of Tarmac Ltd. are fondly remembered. I also send special thanks to the following individuals who supported me over recent years during my visits to England, and provided help in preparation for this book: John Adams, Ivan Jamieson, Derek Harrison and David Porter, formerly of UK Coal; Derek Broughton, formerly of RB (Lincoln) Ltd.; and Rhys Roberts, retired opencaster with experience all over the UK.

Credits for the photographs in this book are indicated at the end of each caption. Those shown as 'KH' were taken by myself. Those with no credit shown are from my collection, which was acquired from numerous sources over the years. Two major additions enhanced this collection: i) while working for Tarmac, I was fortunate to obtain a large number of professional photographs taken at opencast sites of Tarmac and Tarslag; ii) the staff at the Opencast Executive headquarters at Mansfield were very helpful in providing me with a number of surplus photographic prints. I send special thanks to the individuals involved.

Some data and early facts were taken from Peter Grimshaw's book *Sunshine Miners* published in 1992 to celebrate 50 years of British opencast coal-mining; so thanks to Peter for that. Also a 'must read' is the book *Opencast Images 1986–1994* by Dave Wootton, which carries forward from the period covered by this book. It is highly recommended for its accuracy and excellent photography. Last but not least, a special thank you goes to my wife, Barbara, for her love and understanding. She has meticulously edited all my writings over the past 20 years, and in the process has become familiar with the names of most machine manufacturers and contractors that use them!

BRITISH OPENCAST COAL

Introduction

It has been a pleasure to write this book because it allowed me to reflect on my long and rewarding career in earthmoving. From my earliest memories, I was always fascinated by machines that move the earth. Road construction, housing and industrial development, pipelines and opencast coal-mining, all were of great interest to me as a young boy. Living as I did in a coal-mining and industrial area of the English north Midlands, all these types of civil engineering were readily available to observe. By the time I was 11 years old, I was paying regular visits to two opencast coal sites within a short walk from my home in Sheffield. At a young age, I learned what the machines could do, and how to conduct myself safely around them. I also became familiar with manufacturers of the machines and learned their model numbers. The foremen and machine operators were friendly and informative, encouraging me to learn all I could.

At age 13, an appointment was arranged for me to meet the chief engineer of Northern Strip Mining Ltd. (NSM) at their nearby corporate office, and present some drawings and photographs I had taken at their site. He must have been impressed because we discussed earthmoving for well over an hour, and he promised there would be a job for me at NSM when I left school!

One of many early conversations I remember was with a bulldozer operator. He said, 'I know you will be making this a career when you leave school, but you must realise you will have to travel a lot. Construction and earthmoving jobs are never permanent, and when finished you must move to the next one.' Never was a truer word spoken, as after I became qualified as an engineer, I worked on about 15 sites in England and Scotland before moving to Canada for the greater part of my career. Apart from the surface mines for which I had direct responsibility in Canada, I have had the opportunity to travel extensively in North America and visit most of the largest surface mines in coal and oil sands. I have had the privilege of observing the largest machines on earth, thus fulfilling my childhood dreams.

This, my twelfth book on heavy earthmoving machinery, brings me home to my roots in England. I hope you enjoy looking at the nostalgic pictures and reading about the opencast sites and the politics surrounding them. In a book of this size, it is impossible to describe all sites or companies in detail, so I apologise if your favourite is omitted. The sites covered include those on which I had personal involvement or obtained comprehensive data and photographs. They represent a good cross-section of geologically different sites in England, Scotland and Wales, and those employing the most interesting variety of earthmoving machines.

The facts, figures and data covered are taken from my collection of magazine articles, newspaper cuttings, publications by the National Coal Board Opencast Executive and manufacturers' machine brochures and specifications. Also much information was also taken from data I compiled during my many visits to opencast sites and from my own recollection of personal involvement. This literature and recorded data, gathered over more than 60 years, is now referred to as my 'library' and occupies an entire room in my house. It is a valuable resource for my books and articles that have regularly appeared in magazines in Canada, America, Sweden and the UK.

Only brief machine specifications are given on certain machines throughout this book because such detail can be found elsewhere. Because most of this book covers the period before full metrication took hold, specifications used are in standard US (English) weights and measures, including machine weights at 2,000 pounds per ton.

CHAPTER ONE

The Early Years

Mechanized surface excavation for coal, 'opencast' coal-mining, began in Britain in 1941 precisely. Certainly it is believed that surface coal was mined by hand in historic times, but actual excavation using 'modern' earthmoving equipment for mining is accurately documented as starting in 1941 as a wartime measure. During the early years of World War II there was a significant drop in vital coal production from underground mines due to an acute manpower shortage. It was publicly stated that adequate production of food and coal was essential to winning the war, but by early 1941 the situation had become a national emergency.

Members of Parliament considered recalling trained colliers from the armed forces to the pits, but another solution came to the forefront when Albert Braithwaite, Member of Parliament for Yorkshire East Riding and director of contractors Sir Lindsay Parkinson & Co. Ltd., held a private meeting with the secretary for mines. Braithwaite suggested that civil engineering contractors, experienced in operating modern earthmoving equipment, should be given the opportunity to exploit 'surface' or 'outcrop' coal. The government accepted his idea but much work had to be done in an extraordinarily short time.

It was known that high-quality coal existed in many areas within 50 feet of the surface, but unfortunately this coal had not been properly explored and recorded. The Mines Department of the Board of Trade had no surveyed plans for mining here because it was too shallow to extract by underground methods. Although surface mining had been successfully established in America for several decades, this was a new idea for the UK. But, as expected under a national emergency, and with 'wartime' measures in place, activities proceeded speedily, on a schedule that seems unbelievable in today's world of environmental concerns, licences, regulations and public hearings.

In a matter of weeks after the first mention of opencast coal to Parliament, drilling rigs had been sourced and dispatched all over the English and Scottish coal fields to prove and sample coal, necessary entry of land was gained, quantities of coal and overburden calculated, excavation equipment procured (with small capacity at first), coal handling and preparation arranged or plans for new facilities tendered, and contractors' personnel skilled in earthmoving were recruited. Only pure coal seams without containments or partings, and three feet or more in thickness, were considered. The search for shallow coal began in earnest and, within a few months, dozens of suitable sites had been identified with prospect drilling completed.

The first sites

The government plan obtained Treasury approval and work on the first two 'official' sites started in the autumn of 1941. Both were developed and operated by Sir Lindsay Parkinson & Co. Ltd. Land entry was privately arranged through goodwill and agreement with landowners, as statutory powers and existing Defence Regulations did not cover opencast operations.

The first site was Bedgrave Wood near the village of Wales, just south-east of Sheffield. Site work commenced in October, and the first coal production was achieved by 27 December 1941. Production estimates for Bedgrave Wood forecast a total of 200,000 tons of coal to be achieved at a rate of 500 tons per day, and some 200 personnel would be employed.

Some of the first machines on site were a ½-yard Ruston-Bucyrus 17-RB dragline, a ¾-yard Smith 5-20 dragline, a Smith ½-yard shovel, a Caterpillar D7 with a six-yard scraper, five International TD-18 tractors with Euclid 27W dumping trailers, three TD-18 tractors with Ruston-Bucyrus S-90 scrapers and a Caterpillar D8 with a 12-yard scraper. The Smith excavators were manufactured not far away at Rodley, Leeds.

It is interesting to note that three decades later the former Bedgrave Wood site became engulfed by the much larger Meadowgate site operated by Shand Mining, which yielded some 1.75 million tons. In the process of restoring this site, the present Rother Valley Country Park was created.

The second opencast site to be developed was the Orchard site located at the old underground mine of Orchard Colliery Co. Ltd., near Dordon, Atherstone, Warwickshire. Coal was proved to be on this site following a German air raid in June 1941 when bomb craters resulted in surrounding fields littered with lumps of coal! Site preparations began on 3 November 1941, and eventually some 340,000 tons of coal were won, with excavations extended to 70 feet below ground level. Equipment included three draglines, one a two-yard Lima 802 loading into tractor-drawn wagons.

In March 1942, less than a year after Parliamentary approval, Albert Braithwaite was able to report that the first two sites were in production, and that at least 50 million tons of coal within 30 feet of the surface was available for surface mining throughout England and Scotland. Three months later, in June 1942, after only seven months of opencasting, he reported that 40 contractors were working on 22 sites, and a further 63 sites were already in various stages of the planning process.

Supervision of opencast production was placed under the Ministry of Fuel and Power, which was created in June 1942, but that engagement was short-lived. In December of that year it was transferred to the newly formed Directorate of Opencast Coal Production (DOCP), under its first director, Major General Kenelm C. Appleyard. Progress was so rapid that by 1944, 419 sites had been opened up and production had reached 8.6 million tons for that year.

So it is indeed Sir Lindsay Parkinson & Co. Ltd., the company that operated the first two opencast coal sites in Britain, and the driving force of director Albert Braithwaite with his Parliamentary connections, who hold the honour of initiating the opencast coal industry in Great Britain. This achievement is all the more remarkable when, prior to World War II, the British government held no plans whatsoever for future opencast coal production.

Plant and machinery shortage

The rapid increase in opencast production over a few short years resulted in a dire scarcity of heavy plant needed to undertake the immense earthmoving operations at a rate never before seen in the UK. The majority of excavators on the first opencast sites were of ¾-yard capacity or smaller, with only few reaching 2½ yards in size, and scraper capacities ranged only up to 12 cubic yards, clearly insufficient to meet projected coal production.

The acute shortage of equipment became more apparent as the government's demands on the DOCP accelerated. In its first year of operation, the DOCP was tasked with achieving a target of five million tons of coal, and in the following years annual targets of up to 15 million tons were called for. It was obvious that if these outputs stood any chance of success, a new action plan must be achieved. Early in this start-up period, Albert Braithwaite took a visiting American opencast mining engineer onto a UK site. After reviewing the operations, the engineer summed up his comments by saying, 'You are doing it with toys!'

Ken Appleyard, director of the DOCP, visited the USA in 1943 to ask the American government to help by providing extensive fleets of machines, larger than currently employed in the UK, to boost opencast coal production. These machines would be imported under the already established British/American Lend-Lease Programme. Large numbers of American-built tractors, scrapers and other machines had already been supplied under this programme and were working in the UK building air fields for the British and American Air Forces.

Appleyard was successful in securing a promise that all equipment that could be spared would be transferred to the UK. During his visit, he also met Leslie Jones, eastern sales manager for Bucyrus-Erie Company, who recommended three advisors – J. Robert Bazeley, Robert Bailey and Kenneth Youngs – who might be interested in such an assignment. These three gentlemen were steeped in the American surface mining business and subsequently offered their experience and guidance to support the DOCP and the British opencast effort in its early years.

Robert Bazeley was enthusiastic about British opencast coal development and acted as a consultant over a period of many years, making regular visits to the UK accompanied by his wife. He had operated major open pit coal mines in Pennsylvania's anthracite region and, since the 1920s, had employed some of the world's largest excavators. He passed on knowledge gained from his long experience, providing significant input to the initial planning and execution of the UK opencast industry to ensure its future success.

Within a few months, large numbers of second-hand earthmoving machines began to arrive from America and were distributed to opencast contractors. The variety of machines imported encompassed crawler tractors, scrapers, haul trucks, shovels and draglines. But when larger machines such as shovels and draglines started arriving, it soon became apparent that many had almost finished their working lives before leaving the USA.

Manufacturing data, serial numbers and machine condition revealed that some already had worked more than 20 years at mining and construction projects across America. A further problem was that spare parts availability for these American machines was almost non-existent, as their manufacturers had no established distributors in the UK. Once erected and put to work, these tired machines barely lasted the duration of just one site and then had to be scrapped. The situation was aggravated by the tough conditions encountered, often tackling rocky material that should have been blasted. Blasting was ruled out because of cost or proximity to local residents. It was common to see several old broken machines lying idle at the site's 'bone yard' awaiting the scrapper's torch. There were also stories of larger Bucyrus-Monighan draglines being buried on site.

During the 1940s, sites were usually of short duration, most lasting fewer than two years. Machines in use were small by today's standards, but their sheer numbers got the work done. Because the major contractors operated several sites at the same time, they frequently moved machines from site to site in order to complete a special phase of the operation or accelerate local, short-term production. This is in contrast to today's large opencast sites where machines are specified at the start of a project and, if production requirements don't change, remain on site until its completion.

Moving machines from site to site in the 1940s was a relatively easy proposition given their relatively small size. Shovels and draglines were moved on low loaders as a single load with their booms lowered or removed. Crawler tractors up to 128 horsepower (Caterpillar D7 size) size could be carried complete with dozer blades. Scrapers and off-road dump trucks were usually moved in convoy under their own power, with a pilot vehicle leading the procession and a tyre truck at the rear in case of an unplanned flat tyre! Larger machines, of course, had to be dismantled and moved in multiple loads.

In 1944, to further strengthen the mutual cooperation between the British and American opencast industries, the Ministry of Supply formed a six-member mission to tour American opencast coal sites. The group travelled more than 5,000 miles within central and eastern USA, visiting 72 opencast sites and ten manufacturing facilities. The group consisted of DOCP members, directors of opencast contractors and a union leader. Members of the group reported being extremely impressed with what they saw, and the ideas they brought home and implemented led to improved methods and increased efficiency in the British opencast industry.

By the end of 1944, the DOCP reported that 124 excavators and 51 tractors/scrapers had been provided by the American government. These, together with machines provided by British contractors and the UK government, resulted in a total of 621 excavators and 568 tractors/scrapers employed on opencast sites. The DOCP further claimed in 1944 that its opencast coal operations were likely the largest civil engineering job in the world to be run by a single agency. It also boasted the world's largest equipment maintenance organization.

Figure 1.1 Four draglines, three British and one American, are ready to start work at Tarmac's Ponkey opencast coal site, North Wales in 1946. Leading the parade is a British-built 1½-yard Rapier 462, which will be the main stripping machine. This is followed by a pair of Ruston-Bucyrus 19-RBs with ⅝-yard buckets, and bringing up the rear is a Lorain, likely an L-33 or L-37 of similar capacity. The young boy on the bicycle is making acquaintance with the workers near the fuel tank and has come to inspect the site; perhaps a future site foreman or engineer?

Figure 1.2 Ponkey opencast site in full production demonstrates how significant depths can be reached by employing only small machines. The four draglines work as a team, often passing material from one to another, to uncover the coal. Old underground workings are visible here. These are a nuisance in opencast coal production as they must be cleaned out, often using hand labour, and the overall site coal yield is reduced.

Figure 1.3 Tarmac Ltd. fields a multitude of small machines at its Garden Lodge site, Chirk near Llangollen, North Wales in 1945, with tractor-drawn scrapers playing a major part in overburden removal. Caterpillar D8 tractors are pulling LeTourneau scrapers (models LP and YR), probably ex-American Army. The four small draglines in the background are stripping and rehandling overburden and also stockpiling some coal. The far two are ⅝-yard Ruston-Bucyrus 19-RBs, the machine on the left is a ⅝-yard Rapier 423 and the closest machine is a ¾-yard Lorain, likely an L-40 or L-41.

Figure 1.4 Another view of Garden Lodge Site at Chirk, North Wales. Contractor Tarmac is employing two of its Ruston-Bucyrus 19-RB excavators, a shovel and a dragline, to load coal directly into road lorries. At least five men with hand shovels can be seen gleaning every last bit of coal from the footwall or removing lumps of clay or rocks that might otherwise contaminate the coal.

Figure 1.5 An 'army' of scrapers at Tarmac's Garden Lodge site near Chirk, North Wales in 1946. Eleven Caterpillar tractor/LeTourneau scraper outfits can be seen, including 7M and 3T series D7s, and 1H and 8R series D8s. The scrapers are removing topsoil and subsoil layers and storing it in mounds around the site for reuse after mining. The UK has always prided itself in excellent reclamation of opencast sites, unlike America where, prior to the 1960s, topsoil was not normally salvaged.

Figure 1.6 Overburden stripping with tractor-drawn scrapers. The machines are early model Caterpillar D8s with what look like 12-yard capacity LeTourneau model LP scrapers. The tractor on the right is pulling a cable-operated ripper to ease scraper loading in the tough shale. The activity is taking place in 1946 at Vron opencast site near Wrexham, North Wales. The contractor is Tarmac Ltd.

Figure 1.7 A Ruston-Bucyrus 19-RB dragline carefully lifts coal from a thin seam at Tarmac's Vron site near Wrexham, North Wales. This view emphasizes the high mining ratio encountered at some sites. The mining ratio is the number of cubic yards of material that must be moved to yield one ton of coal, and here the high wall appears to yield only one thin seam at its base. Hand labour is much in evidence, as every piece of coal is collected and all visible pieces of rock are removed before the coal is loaded by the dragline.

Figure 1.8 Activities here are at West View opencast coal site in North Wales operated by Tarmac Ltd. At the upper left we see tractors and scrapers reducing the overburden height in preparation for other machines, probably draglines, to remove the remaining overburden down to the first coal seam. A Lima 1201 3½-yard dragline is removing parting material and remaining overburden to uncover the coal. A Ruston-Bucyrus 19-RB shovel is loading coal into a lorry with a wooden-sided body, while the driver supervises the loading from the cab roof to make sure the load is safe for public roads. He may even step into the coal to remove any stray rocks that might otherwise contaminate the load. At right, one of Tarmac's International TD-24 bulldozers widens the haul road.

Figure 1.9 Another view of Tarmac's West View site in North Wales. The Lima 1201 slings about four yards of rock and earth into the previously mined out area of the cut. The International TD-24 bulldozer in the foreground does its part to clear remaining overburden off the coal seam while the coal lorry prepares to back-up left to an awaiting coal shovel (off the picture).

Figure 1.10 Tarmac's Ponkey site in North Wales in 1946 is another example of how relatively deep overburden can be removed with small machines. At left, tractors and scrapers, one pulling a cable-operated ripper, are stripping the upper layer of overburden. These would also be employed to salvage soil ahead of mining and stockpile it for later reuse. In the centre, the exposed coal is being 'cleaned' by hand labour using wheelbarrows and hand shovels. To the right, a team of small draglines is pulling the spoil back to create the necessary room for additional material. They are working in tandem, one passing material to the other.

CHAPTER 1 THE EARLY YEARS   21

Figure 1.11 One of the largest opencast coal sites of the 1940s was Fordbridge Meadows just north of Alfreton, Derbyshire, operated by W.J. Simms, Sons & Cooke Ltd. This scene in 1948 shows a Monighan 6-W walking dragline removing overburden with its 150-foot boom and six-yard bucket. The machine, originally built in America in 1930, was erected at Fordbridge Meadows in 1946 and worked there for five years. After that it was moved three miles to a site at Blackwell, and then moved again to a site near Durham where it worked for a further ten years. A Monighan 6150 walker is seen in the background, while a 19-RB loads coal into a Bedford lorry.

Figure 1.12 Working at Simms' Fordbridge Meadows opencast site in 1948, a ½-yard NCH 150 shovel (left), built by Newton Chambers Company at Thorncliffe near Sheffield, and a ⅝-yard Ruston-Bucyrus 19-RB shovel, built at Lincoln, load coal into Bedford lorries. Advancing the cut in the background is the massive Monighan 8160 walking dragline with its 13-yard bucket.

Figure 1.13 This Monighan 8160 originally worked on levee construction in America from 1931. Once re-erected in 1945 at Townend Farm opencast site at South Normanton, Derbyshire, it claimed to be the largest excavator working in the UK. It carried a 13-yard bucket on a 168-foot boom and weighed some 675 short tons. Tub diameter was 36 feet, overall machine width 52 feet and each walking shoe measured 44 feet by seven feet. This photograph was taken in 1948 after the Monighan moved to Simms' Fordbridge Meadows site (credit: *Art Catino collection*).

Figure 1.14 The Monighan 8160 dragline excavator takes a step at Simms' Fordbridge Meadows site, supervised by the banksman. Only three of this model were built, all in 1931. The model numbering system was intended to indicate the bucket size in cubic yards, followed by the boom length in feet. But in fact, only one of the three actually met these parameters, with customers modifying boom lengths to suit their own operations. Simms referred to this machine as a Monighan 10-W, as that model had similar specifications. A noisy five-cylinder Fairbanks-Morse 33-C-14 engine, developing 450 horsepower at a low 260rpm, powers the machine (credit: *Art Catino collection*).

CHAPTER 1 THE EARLY YEARS 23

Figure 1.15 This Monighan walking dragline, built in 1930, is a prime example of the many used machines imported from America under the USA/UK Lend-Lease Programme. It's a model 6150 operated by Simms at its Fordbridge Meadows site, and carries an eight-yard bucket on a 175-foot Duralumin boom. A large diesel engine provides the power, and the hoist and drag drums operate through air-controlled clutches and brakes. The 6150 was produced from 1929 to 1932 (credit: *Art Catino collection*).

Figure 1.16 One of the oldest models of Monighan walking draglines known to be imported into the UK is this model 3-T. Known as the Martinson Tractor, invented by Oscar Martinson in 1913, the original walking system utilized chains to carry the shoes rather than the improved rigid type shown here. This particular machine has been upgraded to the later Monighan system where the shoes are rigidly attached to the walking spuds. The Monighan 3-T was built from 1913 to 1925 and five are known to have been imported into the UK. A single 3½-T and a 4-T (3½ and four cubic yards capacity) were also imported.

24 CHAPTER 1   BRITISH OPENCAST COAL

Figure 1.17   A.M. Carmichael Ltd. of Edinburgh was a significant player in opencast coal production in Scotland. Eight draglines can be seen here at one of Carmichael's sites of unknown location. Often up to three draglines worked together, passing material from one to another to cast the material back far enough to clear the working area. The draglines in the picture (from left to right) are: Ruston-Bucyrus 54-RB; likely a Ruston-Bucyrus 43-RB with boom down; Ruston-Bucyrus 55-RB; likely another 43-RB in the cut; unidentified machine with only its boom showing; Monighan 6-W walker; and a pair of Lima 1201s pulling back material on the spoil pile.

Figure 1.18   Here's another view of A.M. Carmichael's site showing activities often seen in the 1940s and 1950s, where multiple small machines are employed to uncover relatively deep coal seams. In this case, the deep overburden results in a very narrow cut, and the coal is being loaded by a ⅝-yard Ruston-Bucyrus 19-RB shovel in the foreground. Of course a single larger dragline could have replaced many machines, but they were just not available in the UK when the plan to commence opencast coal in the 1940s was put into action at very short notice.

Figure 1.19 Although mechanization in the form of excavators had arrived in the UK, hand labour was much in evidence when working in very thin coal seams or cleaning out old underground workings. The group of men seen here at one of Dowsett's sites in the north of England is cleaning the coal ready for loading into lorries. Working around the excavators, the labourers separate clay and rocks from the coal to ensure a clean product. The man on top of his lorry is filling voids with a hand shovel and making sure nothing will drop onto public roads on his journey to the disposal point. The lorry in front of him has been 'side-boarded' to maximize its allowable gross vehicle weight (GVW) when carrying light-density coal.

Figure 1.20 Another rare machine in the UK is this Monighan 4-W walking dragline, shown here at an unknown location uncovering coal assisted by a two-yard Lima 802 crawler dragline. The machine shown is one of only three 4-Ws imported into the UK and has been re-engined from original with a Petters SS4 engine. All three were built in the 1920s and came to the UK in 1944. The 4-W carries a four-yard bucket on a 100-foot boom and weighs approximately 210 tons in operation. Its tub (circular base) is 24 feet seven inches in diameter, and each walking shoe measures five feet wide by 31 feet long.

Figure 1.21 Holloway Brothers (London) Ltd. operated this site near Whitley Bay, Northumberland from 1949 to 1951. Six draglines are seen uncovering coal in an orchestrated sequence, each doing its part to expose relatively deep coal, not possible by one single machine. The draglines from left to right, are: A Ruston-Bucyrus 54-RB with 2½ yard bucket; a Rapier 470 (not confirmed); a three-yard Link-Belt K608; a Lima 1201 with 3½-yard bucket; a Lima 802 with standard two-yard bucket; and bringing up the rear another Lima 802. In the foreground, a Ruston-Bucyrus 19-RB on hire from Fen Drains and Excavations Ltd. of Whittlesey loads a pair of Bedford lorries with coal. A Caterpillar D2 bulldozer assisted by the usual hand labour keeps the coal surface clean.

Figure 1.22 Whitley Bay opencast coal site in Northumberland was one of the largest sites in the north east to open in the 1940s, yielding just under a million tons of coal over a two-year period. Contractors Holloway Brothers used a large fleet of machines including more than a dozen excavators, as seen here. In the foreground, a Ruston-Bucyrus 24-RB dragline is pulling the spoil back to make room for the interburden material from between the two seams. This is being excavated by a two-yard Lima 802 dragline and thrown against the spoil pile. Ruston-Bucyrus 19-RB shovels are assigned to coal loading at each coal seam. Further back, a Bucyrus-Monighan 5-W walking dragline is removing the main overburden, while in the distance a team of Lima draglines excavates overburden, assisted by more draglines pulling back on the spoil piles. Topsoil is being salvaged by tractors and scrapers at the left. Established in London in 1882, Holloway Brothers was a major international civil engineering company that completed many famous buildings, bridges and railways in the UK, Australia and the Middle East. In 1964, John Laing plc acquired the company.

Figure 1.23 An aerial shot of Dowsett's Linton opencast site near Widdrington, Northumberland, which was in operation from 1949 to 1951. A three-yard Monighan 3-W walker, just visible in the centre of the cut, is the main stripping machine. Scrapers have already stripped topsoil and subsoil from the area ahead of the cut, and this material can be seen stored in stockpiles around the site for later use in the restoration process. This site yielded 1.4 million tons of coal. Now fully restored, it is interesting to note that most of the surrounding fields seen in the picture became part of much larger opencast sites in the decades that followed. Coal is still mined today around this most prolific area of opencast coal production in the United Kingdom.

Figure 1.24 Starting in the late 1940s, Tarslag operated the Scholes opencast coal site located north-west of Rotherham, Yorkshire, close to what is now the M1 motorway. This busy site employed a large variety of machines. Here, coal is being loaded into Bedford lorries by a pair of Ruston-Bucyrus shovels, a ⅝-yard 19-RB (foreground) and a ⅜-yard 10-RB (centre). Typical of British opencast coal sites are the labourers with hand shovels. Working close to the shovels, and in visual communication with the drivers, the labourers keep the coal clean and free from contaminants, and ensure that every last piece of high-quality black gold goes to market. On the left, the site foreman keeps a watchful eye over the proceedings. In the background a trio of Lima 802 excavators removes overburden.

CHAPTER 1 THE EARLY YEARS 29

Figure 1.25 Another section of Tarslag's Scholes site shows rocky overburden being removed by a Ruston-Bucyrus 54-RB dragline with 2½-yard bucket (background). Having reached the end of the cut, the 54-RB has apparently run out of 'spoil room' and must dump spoil on top of the high wall where the overburden has yet to be removed. This material will be rehandled when the next cut is started. In the centre of the picture, a 1¾-yard Ruston-Bucyrus 43-RB dragline sits on the spoil side and pulls back material to properly expose the coal that has been buried by too much spoil. A ⅞-yard Ruston-Bucyrus 24-RB shovel is assisting the dragline to remove this rocky overburden, passing material over to be hoisted out of the way. In the foreground, a Ruston-Bucyrus 19-RB shovel loads a pair of Bedford lorries.

Figure 1.26 Lots of dragline activity here in 1950 at Tarslag's Scholes site near Rotherham. From left to right, the machines are: a Ruston-Bucyrus 5/8-yard 19-RB pulling back the spoil; a rare Rapier 462 dragline with 1½-yard bucket; a ⅞-yard Ruston-Bucyrus 24-RB; and a 1¾-yard 43-RB dragline. This part of the Scholes site appears far less rocky than other areas and lends itself to excavation by the site's many draglines.

30 CHAPTER 1 BRITISH OPENCAST COAL

Figure 1.27 This trio of Lima 802 draglines is removing overburden at Scholes site about 1952. The shovel in the cut is passing material to the dragline at the right, which would cast it further to the spoil pile. In the distance scrapers can be seen removing and storing topsoil before stripping the overburden. The 802 was built at the Lima Locomotive Works, Lima, Ohio, USA from 1939 to 1954 and sold in the UK by Jack Olding & Co., Hatfield, Herts. The shovel version carried a two-yard dipper on a boom measuring 24 feet six inches long, while the dragline version was offered with booms from 50 to 80 feet long and buckets from 1½ to 2½ cubic yards capacity. The 802 tipped the scales at about 70 tons.

Figure 1.28 At Tarslag's Scholes site, tractor-drawn scrapers are seen here removing the final layer of material to expose the coal. A pair of Caterpillar D8s and a D7 (foreground) are pulling LeTourneau scrapers. At the rear, accompanying the American equipment, a 1½-yard Rapier 462 dragline built at Ipswich, England, removes the bulk of the overburden while another dragline pulls back the dumped material to create more room.

CHAPTER 1 THE EARLY YEARS 31

Figure 1.29 By the end of 1944, 124 excavators were reported shipped from America to assist with the British opencast coal effort. A number of these were manufactured by companies previously unknown in the UK. Consequently, obtaining parts and service support for these machines was extremely difficult, but thanks to some innovative repair work by talented mechanics and locally manufactured parts, the machines remained in service. Examples of these American machines shown here are: (a) Koehring; (b) Lorain; (c) Monighan; (d) Northwest; (e) Manitowoc; (f) Link-Belt.

CHAPTER TWO

The 1950s

By the early 1950s, opencast coal mining in Britain had settled down to a well-organized and profitable industry. The wartime panic to extract coal 'at any cost' was over, and restoration of sites showed vast improvement. By 1950, annual coal production had reached 12.4 million tons with 78 million tons produced since 1941.

In April 1952, at government insistence, the National Coal Board (NCB) reluctantly took over responsibility for opencast production. The NCB's own staff had previously only supervised underground mining and had no experience in bulk excavation with large earthmoving equipment. The Department of Opencast Coal Production, active since 1942, then became the Opencast Executive (OE) of the National Coal Board, and foundations were laid for a profitable organization. As a new department of the NCB, and largely autonomous, the OE supervised opencast contracts undertaken by civil engineering contractors.

The OE's role was to acquire land, carry out site exploration, prepare tender documents and award contracts following competitive tendering. It also introduced restoration regulations and usually employed at least one inspector at each site to ensure contract conditions were met. Contracts were awarded to experienced civil engineering or earthmoving contractors who were free to undertake or modify site design and choose the equipment they needed. The OE would place advance orders well ahead of contract dates for equipment that might take more than two years to deliver. In some cases the OE would supply large expensive items on a rental basis should the duration of the contract be too short for the contractor to justify such an expense. From the outset, opencast mining operations in the UK were – and still are – carried out by private firms experienced in this type of work.

Equipment size increases

During the 1950s earthmoving equipment on opencast coal sites increased in size only marginally, but productivity increased dramatically. This was due to improved working methods and more efficient operations after the OE began to award larger contracts of longer duration. The British/American Lend-Lease Programme wound down, and contractors were able to purchase new equipment from British as well as American manufacturers.

Hindsight reveals that most opencast coal sites in the 1950s could have benefited by using a small number of larger machines instead of a multitude of small machines. But given the many two- or three-year short contracts, and the relatively small surface area of the average site, use of multiple machines was still profitable and far cheaper than underground mining. It was estimated that if overburden could be deposited within a reasonable distance, opencast coal cost approximately a quarter that of coal mined by underground methods.

Use of relatively small diesel-powered machines also provided flexibility as they were able to move easily from area to area, or seam to seam for coal blending. But it involved greater employment numbers, and was less efficient when depth of overburden increased to the extent that material had to be rehandled from one machine to another.

By the end of the decade, some of the world's largest machines were operating in Britain including 50-ton Euclid trucks imported by G. Wimpey & Co. Ltd., but they were few and far between, and worked only on the very largest, long-term sites. In 1959 a mere 14 excavators with capacities greater than eight cubic yards were working on British opencast coal sites (see Table 2.1).

TABLE 2.1 Excavators greater than eight cubic yards capacity operating in the UK in 1959.

Make	Model	Size (yards^3)	Site and Location	Contractor
Rapier	W600	11	Ox-Bow, Yorkshire	Parkinson
Rapier	W600	11	Coney Warren, Yorkshire	Mears Brothers
Marion	7400	12	Maesgwyn, South Wales	Wimpey
Marion	7400	12	Gawthorpe Hall, Lancashire	Wimpey
Marion	7400	12	Kippax, Yorkshire	Parkinson
Marion	7400	12	Dunraven, South Wales	Parkinson
Bucyrus-Monighan	15-W	13	Tramway, Derbyshire	R. McAlpine
Bucyrus-Erie	1150-B	25	Acorn Bank, Northumberland	Costain
Bucyrus-Erie	1150-B	25	Acorn Bank, Northumberland	Costain
Bucyrus-Erie	1150-B	25	Tirpentwys, South Wales	Wilson, Lovatt
Bucyrus-Erie	1150-B	22	Tirpentwys, South Wales	Wilson, Lovatt
Marion	7800	30	Radar North, Northumberland	Crouch
Marion	7800	30	Radar North, Northumberland	Crouch
Krupp	Sch.Rs 250	Bucket wheel	Radar North, Northumberland	Crouch

Although the list is short, it is remarkable that the British opencast coal scene included six of the world's largest draglines in the 1950s. These were the four Bucyrus-Erie 1150-Bs and the two Marion 7800s.

First to arrive near Bedlington, Northumberland, was the 1150-B at Ewart Hill site operated by Sir Lindsay Parkinson & Company. Imported in 1949 as a second-hand machine from America, it was equipped with a 25-yard bucket, 180-foot boom and weighed some 1,200 tons. After this site finished in 1954, the 1150-B walked to the adjacent Acorn Bank site operated by Costain Mining Ltd., where it was joined by another 1150-B dragline of similar specification, freshly imported by the OE for this site. The Acorn Bank site is described in more detail in Chapter 7.

In 1954, the OE imported two more second-hand Bucyrus-Erie 1150-B walking draglines from America in 1954. These worked in the mountainous region of South Wales at the large Tirpentwys opencast site operated by Wilson, Lovatt & Sons Ltd. until 1963. These 1,200-ton machines were equipped with 180-foot and 200-foot booms, and carried 25 and 22 cubic yard buckets respectively.

The arrival of these four 1150-B walking draglines for work in the British opencast coal industry provided the capability to efficiently uncover coal seams at a much lower levels than previously possible. As some of the largest mobile land machines in the world, they were symbols of the industry's confidence and determination to survive and prosper.

Even greater heights were reached in 1955 when the OE purchased two brand new Marion 7800 walking draglines from Marion Power Shovel Company of Marion, Ohio. They were for use by James Miller & Partners at a new site, Radar South, first in a long succession of opencast coal sites within a six mile radius of Widdrington, Northumberland. The two at Radar South were specified to carry 22-yard buckets on booms 280 feet long. Although bucket capacities were slightly less than those on the Bucyrus-Erie 1150-B draglines, the booms on the Marions, some 100 feet longer, gave the latter a much extended working range. Following a boom collapse, the Marions were refitted with shorter booms and 30-yard buckets as shown in Table 2.1. Details of the Radar sites can be found in Chapter 7.

With few sites large enough to accommodate such gigantic machines, medium-sized machines began to appear at relatively small sites of short duration. Where 'electrification' of a site (electric power distribution system) was justified, electric shovels were employed, the first being some second-hand Ruston-Bucyrus 100-RB shovels of 3½ cubic yards capacity. Parkinson purchased two larger five-yard 120-RB shovels in 1950 and 1952, and the OE purchased four more in 1954 and 1955 for distribution to its contractors.

When Ruston-Bucyrus replaced its 100-RB with the 4½-yard 110-RB in 1955, and the 120-RB with the six-yard 150-RB in 1958, these new models became the electric shovels of choice in the UK. Parkinson was the first opencast coal contractor to place both models in service. The 110-RB was also available as a diesel-powered shovel or dragline, which was preferred by contractors working smaller sites where electrification was unjustified.

In 1955 the OE also imported ten six-yard Bucyrus-Erie 150-B electric shovels before the Lincoln-built 150-RB became available in Britain in 1958. These appeared on sites operated by Parkinson (two), Wimpey (four), Taylor Woodrow (two), James Miller and Derek Crouch. All told, some 74 150-B and 150-RB excavators had been placed on British opencast coal sites by the time the last one was delivered to Murphy Brothers in 1978.

Another equally popular machine on British opencast sites was the Lima 2400. This venerable excavator was originally designed by the Lima Locomotive Works Inc. at Lima, Ohio, one of America's leading builders of steam railroad locomotives. Introduced in 1948, the 2400 was first imported into the UK in 1952 by dealer Jack Olding & Company Ltd., Hatfield, Hertfordshire (later SLD Olding). It was a diesel-powered, 220-ton excavator, available as a six or seven-yard dragline on a 120-foot boom, and found immediate acceptance in the UK. The heavy-duty excavator, strongly built, claimed by some to be over-designed, soon earned an excellent reputation as a tough, reliable machine. Its smart, well-proportioned appearance, with raised cab and aesthetically rounded features made it even more appealing!

With its diesel power, the Lima 2400 was more expensive to run, but it found applications on most sites in the UK because of its many advantages:

- Its size and range made it ideally suited for UK surface mines; large enough to be the main stripping machine on smaller sites, small enough to cope with uncovering coal in geologically complicated areas on large sites.
- Its superior mobility, not tethered to a power cable like its electric counterparts, made it ideally suited to work in isolated areas or on sites unsuitable for electrification.
- Its oversized, wide crawler shoes – resulting in lower ground pressure – added to its mobility.
- Moving from site to site could be performed relatively easily as the machine could be dismantled into major components and shipped in about five loads or less including boom and bucket.
- The 2400 entered a UK market already well-established by Jack Olding & Company, distributors for Lima products; large numbers of 802 and 1201 models were already working successfully.
- The big Caterpillar D397 diesel, rated at 473 horsepower, received good parts and service by the network of Caterpillar dealers.

In 1957 G. Wimpey & Company Ltd. boasted in its news magazine that it owned 17 Lima 2400s, and in July 1958 Jack Olding & Company advertised that 61 Lima 2400s were working in the UK. The updated 2400B appeared in 1966, and although few were imported during the coal recession in the 1960s, more arrived during the boom of the 1970s. By the time the last arrived in the UK in 1980, Olding had imported 72 Lima 2400s into the UK.

Private sites

Most major British civil engineering contractors undertook opencast coal projects at varying levels of intensity, but many small private companies also engaged in opencast coal during the energy boom of the 1950s. They operated what were known as 'private sites', licensed but unsupervised by the Opencast Executive of the National Coal Board. These companies operated independently and sought markets for the coal through private arrangement with customers.

This meant that a company operating a 'private' site shouldered entire responsibility from start to finish. This included exploring the site to verify sufficient coal, purchasing or renting the land, meeting all local authority regulations and by-laws, satisfying environmental concerns, planning site layout and earthmoving sequence, selecting and purchasing the plant and equipment, marketing the coal, arranging transport to the customer and reclamation and final sale of the site. This was in contrast to coal-mining contracts awarded by the OE, who had completed most of this work at time of tender.

A good example of a private company specializing in opencast coal-mining was Sheffield-based Northern Strip Mining Ltd. (NSM). In later years, this company became one of the major players on the British opencast coal scene, but its first two decades were involved with 'private sites'. NSM's origin dates back to the late 1940s with relatively small private opencast coal sites in Yorkshire. These included Burcroft at Sharleston near Wakefield, Penn Hill at Cudworth near Barnsley and sites around Mexborough.

From 1956 to 1959, NSM worked its way through a series of small sites less than a mile from the centre of Sheffield, extracting many thousands of tons of high-quality coal. Known as the Clay Wood group, the sites comprised the Granville Road site, Farm Grounds and Belle Vue Field. On completion, these areas were transformed into valuable building land and are occupied today by technical colleges, schools and their associated sports grounds and playing fields.

Starting with the Granville Road site in 1956, NSM purchased several new machines: two Smith 21¾-yard shovels for overburden removal, a Smith 10⅜-yard shovel for coal loading, and five five-ton lorries for transporting overburden (two Bedfords and three Morrises). Three crawler tractors, a Fowler VF and Caterpillar D4 bulldozers for general duty and a Caterpillar D7 working with a LeTourneau LS scraper to salvage and store surface soil. In the 1950s, NSM owned Lima and Ruston-Bucyrus excavators but they were strong supporters of Smith Rodley excavators made locally at Leeds, and most of their fleet consisted of Smith Super 10, 12 and 21 machines, owning 18 by 1959.

To prepare the Granville Road site for development after coal removal, surplus material was hauled in lorries across Granville Road into the adjacent Farm Grounds site, which was also strip-mined for coal starting in late 1956. For this stage of the operation, one of the Smith 21 excavators was converted to a dragline with 50-foot boom, which would normally carry a ½-cubic yard bucket. But NSM added extra counterweight to the machine allowing it to dig with a full ¾-yard bucket. Muir-Hill 10-B dumpers of three cubic yards capacity hauled the overburden.

In 1957, work started in the adjacent Belle Vue field next to Norfolk Park where additional equipment arrived in the form of a two-yard Lima 802 shovel and another D7 dozer. The Lima dug out the initial box cut, and the material was hauled back into the Farm Grounds to fill in the final cut on that site. The same fleet of five-ton tipping lorries were employed, but this time loaded delicately by the much larger two-yard Lima in two short passes. Coal from all the Clay Wood sites was hauled off site to Burnett & Hallamshire's coal sales depot some three miles away.

As the Belle Vue site progressed and the box cut completed, the Lima excavator was converted to a dragline to take the upper level of overburden, and one of the Smith 21 shovels worked the lower level to tackle some tough sandstone. Blasting was not permitted because of the proximity to residences. As the rocky overburden became tougher and too hard for the dragline to handle, the Lima was transformed back to a shovel and was joined by three Euclid R-15 rear dump trucks built in Scotland. With the Lima 802 in the bottom of the cut and the Smith 21 dragline removing the upper softer overburden, Belle Vue field was completed in 1959.

Total coal production from all British opencast sites throughout the 1950s amounted to 119.7 million tons, with annual production remaining steady at just under 12 million tons. This compared with more than 208 million tons produced annually from underground pits during the same period, indicating that only about 6 per cent of British-mined coal came from opencast.

A new reality

With the establishment of the OE in 1952, the British opencast coal scene improved and prospered during the 1950s. The British/American Lend-Lease Programme was discontinued and new, more powerful equipment became available. Contractors gained experience to operate more efficiently, and coal production remained steady at around 12 million tons per year. All looked well for opencast, but problems loomed on the horizon.

Wartime measures were now in the past, and skilled underground coal-miners back in the pits. They began to eye the opencast operations with apprehension.

They saw giant shovels loading coal from seams where the entire seam could be completely extracted efficiently without having to leave a percentage of coal to form underground pillars. They observed huge tonnages leaving sites employing very few men. Tons per man-hour were ridiculously high compared with underground statistics. The miners couldn't compete with that and the miners' unions were not happy.

Other countries, especially America, have always regarded mining as mining, regardless of the method employed, and most large coal companies perform both underground and surface mining. Great Britain on the other hand was one of few countries that separated underground mining from surface mining, and treated these as separate industries. The underground miners therefore viewed opencast as a threat to their jobs, as it was performed by contractors mostly employing non-union labour. Local MPs sided with the underground miners because they were in the majority.

With war measures now in the past, households near opencast sites were no longer prepared to put up with excessive dust, processions of heavy lorries hauling coal from the sites, noisy machines keeping them awake at night and restricted access to former open country. Although temporary, these inconveniences were no longer tolerated and residents protested. The new era of environmentalism, public hearings, increased regulations, and prolonged applications for new sites had arrived!

Figure 2.1 Machines working in 1952 at Tarmac's Royle opencast coal site at Standish, Wigan, Lancashire. A Marion 111-M diesel shovel is loading some Athey Forged Track trailers hauled by a Caterpillar D8 crawler tractor, while another D8 with a pair of Athey trailers approaches the shovel for another load. A Lima 802 dragline distributes spoil ahead of the Marion. The 111-M diesel shovel carries a 4½-yard standard dipper on a 32-foot boom and weighs 127 tons in operation. Stockpiles of surface soil can be seen in the distance.

Figure 2.2 Caterpillar D8 tractor of late 1940s vintage pulls a LeTourneau YR-12 scraper to collect about 12 cubic yards of subsoil at an opencast site in the early 1950s. The outfit is owned by Tarmac Ltd., who likely took advantage of hundreds of 8R and 2U series D8 tractors becoming available to contractors at giveaway prices after World War II. Caterpillar shipped thousands of tractors to Europe during the 1940s for use by the US armed forces and after the war, huge inventories were sold as government surplus.

Figure 2.3 Lots of dragline activity here at Sir Robert McAlpine's Tramway opencast coal site in the early 1950s near Swanwick, Derbyshire. The large dragline on the right is a rare Bucyrus-Monighan 15-W swinging a 13-yard bucket on a 180-foot boom, the only one of this model ever to work in the UK. This electric-powered machine built in 1940 previously worked in Pennsylvania, USA, until 1947 when it was imported into the UK by McAlpine. The 15-W was one of the largest draglines to use clutches and brakes for the hoist and drag functions, rather than employ close-coupled electric motors. The two draglines on the left are gaining spoil room by pulling back material from the coal seam. They are a Bucyrus-Monighan 5-W walker and a Marion 111-M crawler dragline rated at five cubic yards and 4½ cubic yards respectively. This site finished in 1960 when the two Monighans were scrapped.

Figure 2.4 With the winding down of the Lend-Lease Programme in the 1950s, and the import of well-used, worn-out American equipment discontinued, contractors were quick to purchase modern machines. This International TD-24 crawler tractor equipped with Bucyrus-Erie cable blade, purchased new by Tarmac Civil Engineering Ltd., is seen at West View opencast site, North Wales. The TD-24 at 180 flywheel horsepower was one of the largest tractors available at the time.

Figure 2.5 Activities in 1955 at Tan Llan opencast coal site at Coed Talon, Flintshire. Machines include a pair of 3½- yard Lima 1201s removing overburden, a Ruston-Bucyrus 24-RB in the cut and a Ruston-Bucyrus 19-RB loading coal. A fleet of Caterpillar tractors and scrapers reduces the height of the upper bench. The site operated by Tarmac Civil Engineering Ltd.

Figure 2.6 M. Harrison & Co. Ltd. of Leeds was another small contractor who entered the opencast scene in the 1950s. A Lima 1201 dragline is the main overburden excavator at this site near Huddersfield, accompanied by a ¾-yard Smith 5/20 shovel loading coal into a shiny new Bedford lorry. Harrisons achieved a total opencast coal production of 2,390,000 tons.

Figure 2.7 A Rapier 424 excavator works with a drop ball to break up an old airfield to make way for strip-mining. The hardcore would no doubt have been used to build the base of a haul road. The ⅝-yard 424 was built at Ipswich, England from 1954 to 1960 when 107 units had left the factory. Ransomes & Rapier Ltd. did supply a number of walking draglines to the opencast coal industry, but few of its smaller crawler excavators appeared on coal sites.

Figure 2.8 Mary Anne site was near Burnley, Lancashire and was operated by Tarmac in the mid-1950s. This view, taken in 1955, shows a Vickers VR180 crawler tractor pulling an Onions scraper. The 17-ton tractor is powered by a 180-horsepower Rolls-Royce engine, and is part of a fleet of tractors and scrapers seen in the background recovering soil. These include Caterpillar D8, 2U and 8R series tractors with LeTourneau scrapers.

Figure 2.9 A busy scene at Tarmac's High Barracks opencast site near Burnley, Lancashire in 1953. The foreground shows a Lima 1201 with 3½-yard bucket loading a fleet of Caterpillar DW-10 wheel tractors and what appear to be Athey side-dumping trailers. Behind these are Caterpillar D7 and D8 tractors pulling LeTourneau scrapers, and working in the background is a Ruston-Bucyrus 33-RB dragline.

Figure 2.10 There's no doubt who is the proud owner of these machines! The Lima 1201 is loading 15-ton Euclid R-15 dump trucks at an opencast site in the north of England. The Euclid R-15 was one of the first models built by Euclid (GB) Ltd. at their Newhouse, Lanarkshire, factory, which opened in 1950. Working close to the high wall, the Lima shovel with 3½ yard dipper removes some tough overburden to reach coal at a lower level. Although mostly imported from America by dealers Jack Olding & Co. Ltd., the Lima 1201 was also built in Britain by the North British Locomotive Company from 1954 to 1960. The British version was powered by a 240 horsepower, four-cylinder Crossley engine.

Figure 2.11 Shown at Hilcote Hall opencast site near Alfreton, Derbyshire, is one of Wimpey's extensive fleet of Lima 1201s. It swings a 3½-yard Hendrix perforated bucket on an 80-foot boom. The early opencast coal sites were often small, many lasting only two years or less, and employed a single stripping machine such as this Lima. Topsoil and subsoil piles can be seen stored at the left. Such close proximity to residences often resulted in noise complaints in the early years of opencasting.

CHAPTER 2 THE 1950s   43

Figure 2.12 Wimpey's Broughton Moor site in Cumberland was in production for two years from 1958. The comparatively shallow overburden up to 30 feet deep was easily handled by the 7-yard Lima 2400 dragline. A Ruston-Bucyrus 22-RB loads coal into Ford and Commer lorries which have been side-boarded to obtain a full rated load of low density of coal.

Figure 2.13 A Monighan 6-W walking dragline being rebuilt for the Extwistle Hall site near Burnley, Lancashire after sitting idle at an adjacent coal site for several years. Tarmac commenced this contract in 1956 and purchased the American-built 6-W from James Crosby & Co. Ltd. who had completed the earlier site. The 6-W was originally built in the 1920s and arrived in the U.K. in 1944 under the British/American Lend-Lease Programme.

Figure 2.14 Tarmac's Monighan 6-W walking dragline at Extwistle Hall near Burnley in Lancashire. Used in an advertisement for BP's Diesolite, this picture, taken in 1956, shows fuel delivery to the dragline, which carried a four-cylinder two-stroke diesel engine rated at 240 horsepower. The fuel tank held 655 gallons and the slow-revving engine drank 10.5 gallons per hour. The 6-W dug with a six-yard bucket on a 100-foot boom, and its base or tub measured 27 feet seven inches in diameter. Machine weight is listed at 275 tons.

Figure 2.15 In 1955, some three years before the 150-RB was built at Lincoln, England, the OE imported the equivalent Bucyrus-Erie 150-B from America for use by major opencast contractors. Ten such machines were imported before the Ruston-Bucyrus 150-RB became available in 1958. These were operated by James Miller & Partners, Parkinson, Wimpey, Taylor Woodrow and Derek Crouch. Where electric power was available, the six-yard electric shovel proved to be a perfect match for loading the 22- to 35-ton capacity dump trucks popular in the 1950s. This scene shows one of Wimpey's 150-Bs loading a 35-ton Aveling-Barford SN dump truck.

Figure 2.16 This photograph shows the first of ten Bucyrus-Erie 150-B shovels imported from America by the OE. It went to work for James Miller & Partners in 1955, three years before the first British-built 150-RB appeared. The shovel is loading a fleet of Miller's Euclid 22-ton haulers. The five major contractors who employed the ten machines declared them a success, proving the demand for this six-yard electric mining shovel in the UK. Ruston-Bucyrus at Lincoln commenced manufacture of the 150-RB in 1958.

Figure 2.17 One of two Rapier W300 walking draglines working at Black Row opencast site at Heddon-on-the-Wall, Northumberland. They were the first two of this model built by Ransomes & Rapier Ltd. of Ipswich, and were delivered in 1957 and 1958. They carried seven-yard buckets on 140-foot booms. Black Row and the adjacent Bays Leap site, where the W300s also worked, were operated by Sir Lindsay Parkinson.

46 CHAPTER 2 BRITISH OPENCAST COAL

Figure 2.18 A Rapier W150 walking dragline carefully dumps a five-cubic yard bucket on a Euclid 22-ton dump truck at Parkinson's Black Row opencast site. For obvious safety reasons, it is standard practice for the dump truck driver to stand well clear of his vehicle during loading by a dragline.

Figure 2.19 One of Parkinson's Ruston-Bucyrus 110-RB shovels loads Euclid R-22 dump trucks at Black Row opencast coal site in Northumberland. First built in Britain in 1955, the 4½-cubic yard excavator was available as a shovel or dragline. Most were electric powered, like this one, receiving power via trailing cable, but some were diesel-electric machines with Caterpillar power.

CHAPTER 2 THE 1950s 47

Figure 2.20 This Bucyrus-Erie 1150-B dragline with 25-yard bucket and 180-foot boom was the first large dragline to operate in the British opencast coal industry. Originally built in 1946, it first worked in the Pennsylvania anthracite field before being imported to the UK in 1949. The picture shows the 1150-B working at Parkinson's Ewart Hill Opencast Coal Site near Bedlington, Northumberland, under the name Parkinson Strip Mining Company Ltd. It remained in Northumberland for the rest of its working days until 1984, spending time with Crouch at Coldrife, Ladyburn, Radcliffe and Togston coal sites.

Figure 2.21 A busy scene in the cut at Parkinson's Ewart Hill opencast coal site near Bedlington, Northumberland in 1952. The 1150-B dragline in the background is stripping overburden with its 25-yard bucket. Coaling shovels include two Ruston-Bucyrus ⅝-yard 19-RBs closest to the dragline, an unidentified Lima shovel and, in the foreground, a ⅞-yard Ruston-Bucyrus 24-RB. The usual labourers can be seen in mid-distance cleaning coal by hand, assisted by a Caterpillar bulldozer. Coal lorries are identified as Fordson and Bedford, with extended sides to ensure a full load of low-density coal.

Figure 2.22 This is one of two Bucyrus-Erie 1150-B walking draglines the Opencast Executive imported second-hand from America to work on the large Tirpentwys opencast site in South Wales. These 1,200-ton machines were equipped with 180-foot and 200-foot booms, and carried 27 or 25 cubic yard buckets respectively. These, and the two similar 1150-Bs at Ewart Hill and Acorn Bank sites, were the largest machines to work in British opencast coal at that time, and symbols of the industry's confidence to survive and prosper. Opencast contractor at Tirpentwys was Wilson, Lovatt & Sons Ltd., whose site produced 1.8 million tons of coal from 1950 to 1963.

Figure 2.23 This fleet of Allis-Chalmers HD-19 and HD-20 crawler tractors with Onions scrapers is stripping topsoil and subsoil prior to overburden removal. The soil will be stored in mounds around the site until replacement on the re-contoured overburden after mining or, where possible, directly placed on reclaimed land at a completed part of the site. From the very early days, the OE insisted on first-class restoration results. Adequate soil placement, proper drainage and soil nutrients were mandatory for successful completion. The big Allis-Chalmers HD-19 crawler broke the world record for size in 1947, weighing some 20 tons without blade, and developing 163 flywheel horsepower with torque converter drive.

CHAPTER 2   THE 1950s   **49**

Figure 2.24 The Lima 2400 proved ideal for British opencast mining conditions. Its six- to eight-cubic yard size was a good match for the 22- to 45-ton dump trucks, and its diesel power afforded maximum mobility on relatively short-term sites. Lima distributor Jack Olding & Co. Ltd. (later SLD Olding) imported 72 Lima 2400 and 2400B machines from 1952 to 1980. The 2400 was powered by a Caterpillar D397 V12 diesel engine rated at 473 horsepower. This 2400 loads a Euclid R-22 rear dump hauler for Taylor Woodrow.

Figure 2.25 Sir Lindsay Parkinson & Co. Ltd. pioneered opencast coal in Great Britain in 1941. It soon commanded sites large and small in every coal mining area. A typical application here, near Rotherham in Yorkshire, shows a 2½-yard Ruston-Bucyrus 54-RB dragline as the main stripping machine, while a Caterpillar 933 Traxcavator, assisted by hand labour, sorts out coal from rock. Mobile machines, such as wheel tractors, small bulldozers and crawler loaders like this one, became available by the late 1950s significantly reducing the need for hand labour.

Figure 2.26 Methley Park near Leeds, Yorkshire, was another of Parkinson's opencast sites of the early 1950s. Coal is being loaded into lorries by one of three ¾-yard NCK 304 shovels at the site. Stripping the overburden is a 2½-yard Rapier 490 crawler dragline, a rare machine in opencast coal. It seems strange that Ransomes & Rapier Ltd. of Ipswich built a full range of diesel and electric shovels and walking draglines, yet did not play a more active role in supplying equipment to the opencast coal industry, especially in the early days when so many machines were imported from America.

Figure 2.27 More activities at Parkinson's Methley Park site near Leeds. A Bucyrus-Monighan 5-W walking dragline with five-yard bucket uncovers the coal, and behind it is a 2½-yard Ransomes & Rapier 490 crawler dragline. The coal is being loaded by a pair of NCK 304 ¾-yard shovels built by Newton-Chambers & Co. Ltd. a few miles down the road at Thorncliffe near Sheffield. Most interesting are a pair of ex-army tanks that have joined the tractor/scraper fleet. They have been rigged for scraper duty by adding a rear-mounted double drum winch to control the scrapers. Such outfits reflect the dire need to uncover coal immediately after World War II, and the extreme shortage of equipment.

Figure 2.28 In 1959 Caterpillar launched its first wheel loader, the 944A, and Parkinson was quick off the mark to purchase one of the first to arrive in the UK. Stockpiling coal from opencast sites was necessary to balance varying production rates with varying delivery rates. It was also necessary to blend certain coals to achieve a specific coal specification. Wheel loaders efficiently handled the various coal stockpiles, and the 944A with its 1¼-yard standard bucket was an ideal size to load licensed road lorries.

CHAPTER 2 BRITISH OPENCAST COAL

Figure 2.29 This Caterpillar 933 Traxcavator assists an NCK shovel to load coal at one of Parkinson's sites near Rotherham, Yorkshire in the late 1950s. Small machines like this crawler loader were invaluable for cleaning the coal surface before loading, and also for removing waste material from old underground workings to leave uncontaminated coal. Such machines greatly reduced the need for hand labour but did not eliminate it entirely. The advent of hydraulic excavators on coal sites a decade or so later, would reduce hand labour to an absolute minimum.

Figure 2.30 In 1960, Mears Brothers (Contractors) Ltd. purchased a brand new Rapier W600 walking dragline and put it to work at Pool Covert site near the river Ouse, Yorkshire. The manufacturer's personnel in white overalls add a touch of professionalism during the machine's final testing. This W600 initially carried a ten-yard bucket on a 186-foot boom and was the last of only four draglines of this model to be produced by Ransomes & Rapier Ltd. It spent its entire life in Yorkshire, moving to about four other sites in the vicinity. Another W600 was operated by Parkinson at the nearby Ox-Bow site.

Figure 2.31 Looking like a giant slice from a knife, Northern Strip Mining's Granville Road site (centre) near the centre of Sheffield is completed in this 1957 aerial photograph. Next to it, the larger Farm Grounds site is still active, while Belle Vue field (lower right) is yet to be started. Knowle House (lower left) was NSM's head office in the 1950s and 1960s. Sheffield Midland station and city centre are seen in the background.

Figure 2.32 This image, taken from the ground a few years after restoration, shows the same area covered in the previous aerial shot (Figure 2.31). Looking from Belle Vue field, the Farm Grounds and Granville Road technical colleges are seen. *KH*

54 CHAPTER 2 BRITISH OPENCAST COAL

Figure 2.33 The 75-ton Lima 802 shovel with its two-yard dipper delicately loads a Morris five-ton lorry in two passes during the box cut phase of Belle Vue site. This material was hauled to the adjacent previously worked out Farm Grounds site. *KH*

Figure 2.34 A Smith 21 shovel tackles rocky conditions at Northern Strip Mining's Belle Vue site, one of the Clay Wood group of sites located less than one mile from the centre of Sheffield. Morris five-ton lorries and three-yard Muir-Hill 10-B dumpers are the haulers. *KH*

CHAPTER 2 THE 1950s **55**

Figure 2.35 Smith 10 and Smith 12 excavators load coal and remove parting material from the main cut at Belle Vue. 'Parting' is a term used to describe thin beds of clay or other contaminants found within a coal seam. 'Interburden', another often-used mining term, is a thicker layer of material between two separate coal seams. *KH*

Figure 2.36 A Lima 802 shovel excavates rocky overburden and loads Euclid R-15 dump trucks at NSM's Belle Vue site. The Smith 12 coal shovel in the foreground, owned by H. Shaw & Son (Mining) Ltd., one of the NSM group of companies, waits to load the coal seam. A labourer cleans the coal surface by hand. *KH*

56 CHAPTER 2 BRITISH OPENCAST COAL

Figure 2.37 Overburden at Belle Vue reached a maximum of no more than 50 feet. At the shallow end, this Smith Super 10 loads coal into a Leyland Comet lorry owned by Burnett & Hallamshire Ltd. It will haul the coal some three miles off site to the company's coal distribution centre. *KH*

Figure 2.38 NSM's Lima 802 dragline strips the upper overburden at Belle Vue site. Belle Vue House is in the background. This machine, popular on British opencast coal sites, carries a two-yard bucket on an 70-foot boom. Power comes from a Caterpillar D17000 diesel putting out 180 horsepower. *KH*

Figure 2.39 Belle Vue field a few years after restoration showing a rebuilt Belle Vue House, now used as a training college. *KH*

Figure 2.40 Less than half a mile from the centre of Sheffield, contractors W. Malthouse Ltd. worked the Park Hill opencast coal site in the mid-1950s on a hill known locally as Skye Edge. The site was notorious for its hard rock, and – being completely surrounded by residential properties – no blasting was allowed. This resulted in tough assignments for the excavators, which caused frequent breakdowns. In the foreground, a new 1½-yard Ruston-Bucyrus 38-RB shovel tackles the rock, assisted by a ¾-yard Smith 21 above. At mid-distance another Smith 21 shovel removes upper overburden, while in the background old Lima and Ruston-Bucyrus shovels are parked after breakdown. *KH*

CHAPTER THREE

The 1960s

At the beginning of the 1960s, future prospects for opencast coal mining in Great Britain looked grim; it seemed almost everyone was against it. Noise and dust from the workings, increased traffic on neighbouring roads and perceived destruction of land all influenced the industry's poor reputation. Most prominent on the list of complainants was the National Union of Mineworkers (NUM), who feared mass unemployment if efficient mining of opencast coal by private contractors was allowed to expand. The NUM was supported by local Members of Parliament who fought long and hard to prevent new sites from opening.

Responding to such pressure, and the fact that coal stockpiles were at high levels, the chairman of the National Coal Board (NCB), Sir James Bowman, announced in 1960 a proposed drastic contraction of the opencast industry. An article in a national newspaper headlined 'All open-cast pits may go' reported his announcement that opencast production would drop from 11 million tons in 1960 to only 7 million tons in 1961, and that opencast would end by 1965 except for 'few long-term contracts' (see Figure 3.1). In addition, overall British coal production would reduce from 250 million to 200 million tons over the five-year period.

Of course we know that opencast did not stop in 1965, but production did drop from a high of 14 million tons in 1958 to an average of some seven million tons per year for the decade beginning in 1961. Nevertheless, the 1960 announcement came as a severe blow to the opencast industry, just at a time when opencast contractors were gearing up for larger contracts. Personnel had gained valuable experience and enjoyed success over the previous two decades, with some spending their entire careers in opencast mining. Civil engineering contractors had discarded obsolete machinery and were well-equipped with modern efficient machines.

In the wake of the 1960 announcement, contractors had to rethink their strategy and decide what to do with pending surplus equipment. Some abandoned opencast coal altogether, while some were fortunate to deploy their equipment on other earthmoving projects such as the motorway construction programme, which was fortunately moving into top gear at that time. Those serious about opencast coal vigorously pursued the few contracts still offered by the Opencast Executive (OE) of the NCB. Big 'long-term' sites unaffected by the cutbacks included Wimpey's Maesgwyn and Parkinson's Dunraven in South Wales, Parkinson's Ox-Bow in Yorkshire, Derek Crouch's Radar North group in Northumberland and Costain's Westfield in Scotland.

Northern Strip Mining Ltd. (NSM) of Sheffield, still a private opencast coal contractor in the early 1960s, was able to open a civil engineering department and take on a variety of contracts including road construction, sewerage construction, reclamation sites, metal and slag recovery, canal dredging and large industrial basement excavations. The largest excavator to appear in the centre of Sheffield at that time, a two-yard Lima 802 shovel from NSM's fleet, helped dig foundations for a Sheffield University extension and the foundations of Sheffield College of Technology.

Another type of project was industrial development in conjunction with coal recovery. These sites appealed to local authorities because land improvement could be gained at no cost to taxpayers. Industrial dereliction, in many cases including surface facilities of old underground coal mines or old factories, would be cleared and replaced by public parks, sports grounds or provide valuable land for housing or new industrial development. Usually of short duration, these sites passed through the approval phases with little objection.

In the 1960s the British motorway programme was underway and this also resulted in some opencast coal work in certain areas. For example, in Derbyshire, the route of the M1 motorway passed over former shallow underground coal workings that would have made the motorway unstable due to potential subsidence. Extraction of all remaining coal between the old tunnels was therefore essential, so the old workings, perhaps excavated more than 100 years before, were exposed by opencast operation and the remaining coal recovered. The area was then backfilled and compacted to Ministry of Transport specifications to ensure long-term stability.

Improved standards

The 1960 announcement, with its subsequent reduction in opencast output, was a wake-up call for the industry and its contractors. It was obvious that future sites would be impossible without public involvement from the early stages of planning through to completion of the projects. Since that time, standards of site operation and restoration have been constantly revised and upgraded, taking into account local conditions and considering all concerns raised by the public. These have been properly addressed and minimized as far as possible. The effects on the environment by every activity in all phases of an operation have been scrutinized and problems mitigated. Examples of some of the hundreds of improvements, many in regulatory form, include:

- Improved soil handling standards, already high when compared with other countries, would include site-specific studies to determine optimal depths of topsoil and subsoil. (Early American surface mines did not salvage topsoil.)
- Site-specific tests to obtain appropriate fertilization, seeding and planting procedures.
- Landscape design to ensure proper drainage and increase cultivatable land.
- Specialized equipment introduced to transplant fully-mature trees from working areas to reclaimed areas.
- Mandatory construction of settlement ponds to prevent any contaminated mine runoff entering streams or rivers.
- Dust reduction achieved by:
  - more frequent watering of haul roads;
  - water jets played on excavation faces;
  - introduction of water mist sprays downwind of sites;
  - temporary or early seeding of soil piles.
- Noise reduction achieved by:
  - advanced highly-efficient silencers fitted to noisy diesel-powered machines;
  - soil piles arranged around the perimeter of sites in the form of baffles to deflect noise and improve visual appearance;
  - restriction of working hours, in some cases avoiding night shifts.
- Improved blasting techniques and restriction to day shifts only.
- Elimination of mud on adjacent public roads by introducing automated wheel-washers for traffic leaving a site.
- More attention paid to securing coal loads before lorries entered public roads, and having lorries leave a site on a timed schedule basis to avoid 'convoys'.
- Safety of workers and public made 'number one' priority. In most cases sites to be completely fenced to discourage trespassers; footpaths and bridleways would be maintained or diverted.

Considering these improved environmental standards, and the described benefits of opencast mining, coupled with the need for anthracite coal found only in South Wales, most of it only mineable by opencast methods, the government soon realized that its proposal to permanently shut down opencast coal made no sense. After all, opencast coal was always profitable, and these profits supported certain underground coal mines that were unprofitable, thus keeping underground miners employed. Also opencast coal was sometimes of a higher quality than coal from underground mines, but when blended together, a saleable product resulted.

As early as 1962, the Minister of Power, Richard Wood, responding to opposition questions, refused to order the NCB to reduce its opencast coal operations or to limit them to anthracite production. He said: 'Opencast operations could help to maintain a profitable market for coal, and the Board should be free to continue them unless there are over-riding objections on agricultural, amenity or general social grounds.'

As coal stockpiles diminished, Henry Collins, production manager for the NCB, announced in late 1962 that the Board was building up a list of potential opencast coal sites to be developed 'when the need arises'. He said: 'To win new customers for coal and successfully compete against oil, we must be able to act quickly.' As a result, opencast coal production did continue throughout the 1960s, although the annual average yield was only approximately half the 14 million tons achieved in 1958.

Some contractors active in opencast coal in the 1960s

George Wimpey & Company Ltd.

George Wimpey & Company Ltd. was the number one opencast coal contractor in the UK in terms of production. In the period 1942 to 1974, Wimpey produced more than 37 million tons of coal from sites in all coal-mining areas of the UK. Wimpey was a leader in mechanized earthmoving and led the way with modern, innovative equipment, often purchasing first-of-a-kind machines for its operations in the UK. As early as the 1950s, the company was running huge tandem-drive (6 × 4) Euclid rear dump trucks of 40 tons capacity imported from America, the largest in the UK at that time. It employed electric shovels and walking draglines purchased directly instead of renting from the OE.

One exception to this occurred in 1961 when Wimpey put to work a British-built Rapier W1800 walking dragline of 40 cubic yards capacity, at that time the largest dragline in the world. Owned by the OE, it was placed in service at Maesgwyn Cap site in South Wales, one of the largest and longest-running sites in the UK. Another first for Maesgwyn was the arrival of three nine-yard Bucyrus-Erie 190-B electric shovels in 1964 and 1965. These machines, imported directly by Wimpey from America, were some 50 per cent greater in capacity than the largest shovels then available in the UK. A detailed description of the Maesgwyn site can be found in Chapter 7.

Wimpey's equipment replacement policy, not often seen in smaller contractors, was that it sold used machines long before their economic life had ended. Keeping only modern machines in its fleet resulted in a higher availability. When sold, mostly to overseas customers, these used machines in good condition with a long working life still ahead would yield a higher price. Former Wimpey-owned electric shovels and two Marion 7400 draglines sold to Canada were reported to be still working some 15 years later.

Sir Lindsay Parkinson & Company Ltd.

Sir Lindsay Parkinson & Company Ltd., pioneer of the British opencast coal industry, was the second largest opencast contractor in the UK. It owned an extensive fleet of earthmoving machines of all sizes, and operated sites at all coal-mining areas in the UK. Its fleet of walking draglines ranged from the small Monighan machines obtained through the British/American Lend-Lease programme in the 1940s, to the giant 1,200-ton Bucyrus-Erie 1150-B, one of the largest excavators in Europe, imported for the Ewart Hill site in Northumberland.

But Parkinson did not totally rely on machines imported from America. The company supported British industry by purchasing many Ruston-Bucyrus machines built at Lincoln. These included significant numbers of 110-RB and 150-RB electric shovels, two 3-W and six 5-W walking draglines. From Ipswich-based Ransomes & Rapier Ltd. it purchased four five-yard W150, two seven-yard W300 and one 11-yard W600 walking draglines. With the exception of the W600, these machines were dismantled several times and moved to new sites.

The two Rapier W300 draglines began life in Northumberland and served at Black Row and Bays Leap sites near the village of Heddon-on-the-Wall. Built in 1956–1957, they were two of only four W300s built, but were of advanced design, constructed in modular units bolted together on-site. The modular units were transported complete with machinery already installed, and mechanical and electrical connections ready for quick assembly. This allowed machines to be moved and re-erected in a matter of weeks instead of months. This modular design became popular with other dragline manufacturers in the 1970s.

After the Northumberland sites, Parkinson's W300s spent time at Ox-Bow and Anglers sites in Yorkshire, and Pen Gosto in South Wales. They were finally scrapped in 1984.

The W300 appears a bulky and heavy machine for one rated at only seven cubic yards capacity. Its working weight with standard 140-foot boom was 420 tons. This compares with 218 tons for the Lima 2400 of similar capacity, although with a shorter boom at 120 feet. Much of the extra weight of the Rapier would be due to additional steel required to form the rigid modules, and the extra 20 feet of boom would be an advantage when spoil had to be cast well clear of the coal to avoid spoil sloughing.

One of Parkinson's largest sites was Ox-Bow located at Templenewsam, just south east of Leeds, Yorkshire. Beginning in 1957, this site eventually encompassed three additional adjacent sites (Charcoal, Gamblethorpe and Skelton) and provided Parkinson with opencast work in the area for nearly 40 years. The sites employed a variety of equipment including a dragline converted to a crane to hoist coal in large skips from the deep cut. The Ox-Bow site is described in more detail in Chapter 7.

Another large site operated by Parkinson was the long-running Dunraven site near the village of Rhigos above the Vale of Neath, South Wales. Here Parkinson employed one of its Marion 7400 draglines, obtained second-hand from Sweden. Other machines included a Marion 191-M with a 14-yard dipper, a seven-yard Lima 2400 dragline, a Ruston-Bucyrus 110-RB dragline and Bucyrus-Erie 150-B and 190-B shovels with six- and nine-cubic yard dippers respectively. These were imported from America before the similar British 'RB' models became available.

Derek Crouch (Contractors) Ltd.

Derek Crouch (Contractors) Ltd. was the third largest opencast coal contractor in the UK, bolstered during the 1960s by its successful capture of a majority of the contracts in north-east England, known as the Radar North group. This area on the Northumberland coast contained huge quantities of high-quality coal, with relatively flat thick seams found at reasonable depth. This enabled opencast mining to be carried out on a grand scale, not normally possible in the UK.

For Radar North in 1960, Derek Crouch imported from Germany a Krupp bucket wheel excavator, the only machine of this type ever to work in opencast coal in the UK. With its associated conveyor system and mobile spreader on the spoil area, the excavator removed a thick layer of glacial overburden. As the 1960s decade progressed, Crouch obtained adjoining sites Coldrife, Radcliffe and Ladyburn, and for these contracts employed five of the largest walking draglines in the UK, as well as several large electric shovels. Crouch also took ownership of the largest dragline in Europe, the 65-yard Bucyrus-Erie 1550-W named 'Big Geordie', which started work in 1969. Full details of the Radar North sites are given in Chapter 7.

In 1964, Derek Crouch was awarded the large Abercrave opencast site east of Swansea, South Wales. With its opencast activities already well-established in the north-east of England, the South Wales venture represented a major expansion for Crouch. The new site employed one of the 1,200-ton Bucyrus-Erie 1150-B draglines moved from the former Terpentwys site, a smaller Rapier W150 dragline and a fleet of shovels and trucks. The Abercrave contract called for the delivery of 2.3 million tons of coal over a nine-year period, and with complicated geology and overburden-to-coal ratio varying between 25 and 30 to one, the contract was a significant undertaking. More details of the Abercrave site are found in Chapter 7.

W.J. Simms Sons & Cooke Ltd.

W.J. Simms Sons & Cooke Ltd., general civil engineering and building contractor of Nottingham, was not one of Britain's major opencast coal players, but always had one or two sites running from the very early days of opencasting until it wound down this work in the mid-1980s. As described in Chapter 1, Simms operated some giant Monighan walking draglines in the 1940s, the largest excavators in the UK at that time. After these draglines were scrapped, Simms continued with a succession of relatively small-output sites in Derbyshire, such as Wrang Farm, Doe Lea and Bonds Yard near Chesterfield, Sunnyside at Tibshelf and Shortwood at Trowell, Nottinghamshire. Doe Lea was one of several short-term opencast sites on the alignment of the M1 motorway in Derbyshire where unstable, underground mine workings were excavated and then backfilled with material compacted to Ministry of Transport specifications to ensure long-term stability for the new road.

The larger Shortwood site was mainly a dragline job employing a Ruston-Bucyrus 43-RB, two 54-RBs and a diesel-electric 110-RB, the latter on hire from Shephard Hill & Company. The 168-acre site also featured a variety of scrapers including seven-yard LeTourneau model D Tournapulls and a collection of older model Caterpillar D7 and D8 crawler tractors with pull-type LeTourneau and Caterpillar scrapers. The site yielded some 300,000 tons of coal, and part of it also extended across the alignment of the M1 motorway. The Trowell Service Station is located on this former opencast site.

In 1980, Simms branched out bravely in a big way by importing five American P&H 1200 hydraulic excavators for coal contracts at Amberswood and Miller's Lane sites near Wigan, Lancashire and Pica near Whitehaven, Cumberland. But these 13-cubic yard excavators did not perform to expectations and were discontinued by P&H just a few years after their introduction. This may have contributed to Simms' decision to withdraw from opencast coal contracting by the mid-1980s.

Martin Cowley Ltd.

Martin Cowley Ltd., based at Clay Cross, Derbyshire, started business in 1947 as a small earthmoving contractor. It soon moved into the lucrative opencast coal business in the 1950s and completed several small sites around Derbyshire and Nottinghamshire. In the early 1960s it picked up a number of small opencast sites in the path of the M1 motorway to be constructed a few years later.

At the same time, Cowley took on a number of civil engineering contracts including the Stevenage by-pass on the A1 trunk road, drainage and sewerage schemes at Ramsgate and West Bridgford and a major reservoir at Leek, Staffordshire. Cowley also established a London office to deal with overseas contracts. Most notable of these was a £10 million contract won in 1963 against fierce international competition for the restoration of the Hejaz Railway in Saudi Arabia. The three-year contract involved restoring a 525-mile length of railway that ran from Damascus in Syria to Medina in Saudi Arabia, destroyed by T.E. Lawrence (Lawrence of Arabia) during World War I.

In 1958, the company undertook one of the NCB's most significant opencast coal contracts in Derbyshire. The contract specified the extraction of an estimated 989,000 tons of coal from the Sud site at Barlow near Chesterfield. Although coal tonnage was not exceptionally large by today's standards, it took well into the 1960s to complete the coal extraction and reclaim the site, much longer than intended. More coal was found than the contracted amount and before the job was completed, main contractor Cowley was bankrupt. A second contractor was brought in to complete the restoration. Cowley's Hejaz railway contract was never completed and the line lies derelict today.

Sud, however, was not the major contributing factor in Cowley's failure; poor financial management was the main reason stated in the bankruptcy hearings. Nevertheless, the high coal-to-overburden ratio and tough rocky overburden material at Sud aggravated the situation, and any profits from the site were not enough to save the company.

Main overburden removal equipment at the Sud site was a fleet of four Lima excavators: three six-yard 2400s and a 3½-yard 1201. These initially worked in conjunction with a large fleet of 22-ton Euclid R-22 haul trucks and formed a series of benches extending from one edge of the site to the other. These uncovered multiple thin seams of coal, with the lower seam uncovered by the Lima 1201 equipped as a dragline that cast the excavated material into the previously mined-out cut. The haul trucks travelled on prepared haul roads across the active pit at both sides of the site, and dumped into the worked-out pit, eventually reaching original surface elevation.

Rocky overburden was drilled by lorry-mounted Reichdrills and blasted using ammonium nitrate. Blasting was often minimized because of its high cost, resulting in massive boulders to be handled by the Lima shovels and Euclid trucks. Often grossly overloaded, these heavy-duty machines stood up to the task, even where the Euclids were called upon to negotiate steep inclines on their way to the dump. As the site moved progressively into deeper overburden, the number of benches increased. Also one of the larger 2400s was converted to a dragline to replace the 1201 in the bottom of the pit. The 2400 was changed from dragline to shovel and back to dragline during the course of every new cut.

Ruston-Bucyrus 19-RB and 22-RB shovels loaded the coal into road lorries, which hauled the coal to a local

NCB disposal point as there was no washery on site. Caterpillar D8 dozers maintained haul roads, shovel and dump areas, while road maintenance was handled by Aveling-Barford 99H motor graders and a rare Blaw-Knox BK-12 grader.

In the early 1960s, Cowley replaced some of the ageing Euclid R-22 haul trucks by new Euclid R-27 models of 27-ton capacity. Then in late 1964, the company decided to replace the remaining R-22 dump trucks with a fleet of new 45-ton Euclid R-45s. At this stage, almost seven years after start-up, the site was already in the restoration stage with coal extraction almost completed. One would have to wonder at the soundness of this decision given how precarious the company's financial situation must have been at the time. A year later the company was in receivership.

Backfilling the final void at Sud was a huge task. Two of the Lima 2400 shovels were moved to the top of the dump and worked their way down to the elevation prescribed in the plan. It was also a long haul for the trucks to cross the site and backfill the final cut, the deepest on site. And all this after coal deliveries had come to an end. Cowley was unable to finish the job; it was completed by another earthmoving contractor after legal details were worked out.

Sir John Jackson Ltd.

Sir John Jackson Ltd., the reputable British civil engineering contractor established in 1876, ventured into opencast coal-mining in the late 1940s. With a long history of mechanized earthmoving equipment, including operation of large steam shovels on overseas contracts in the 1920s, its management was more than qualified to enter the opencast coal business in the UK. Some of Sir John Jackson's earlier achievements include construction of docks at Middlesbrough, Hartlepool, Plymouth, Swansea and Sunderland, the harbour at Dover, a section of the Manchester Ship Canal and the foundations of Tower Bridge, London. Overseas work included harbours in Singapore and South Africa, irrigation work in Mesopotamia and Lebanon, and a railway in Bolivia.

The company's opencast coal sites were mainly centred in South Wales, and the firm became well-known around the villages of Trimsaran, Pont Henry, Carway and Five Roads in Carmarthenshire. The Carway House site was one of the company's largest, employing an assortment of equipment including three Lima 2400 excavators, one as a shovel, two equipped as draglines.

For unspecified reasons, the company reduced coal production in the early 1960s and had withdrawn altogether from opencast coal by the mid-1960s. Its last site was Penllergaer near Pontardawe in Glamorganshire. The company achieved a total coal production of three million tons during the course of its opencast activities.

R.A. Davis (Midlands) Ltd.

R.A. Davis (Midlands) Ltd. was another contractor whose name became prominent in the opencast field. One of its most notable sites was Poplars located near Cannock, Staffordshire, one of the longest-running sites in the Midlands. Beginning in 1957, the site boasted a diverse fleet of equipment including many on test by their manufacturers or sales agents. The site's early days saw Ruston-Bucyrus 43-RB shovels loading overburden into Scammell Mountaineer dump trucks augmented by fleets of tractors and scrapers, and Ruston-Bucyrus 19-RB and 22-RB shovels loading coal.

By 1960 the site was home to at least four Lima 2400 6-yard shovels and a Marion 4½-yard 111-M dragline. R.A. Davis fielded a 30-strong fleet of Aveling-Barford SN 35-ton dump trucks, and was also a big customer of AEC dump trucks, especially the 15-ton 6 × 4 type. AEC, through its distributor Scottish Land Development Corporation, carried out considerable development work on AEC 23-ton and 27-ton dump truck models at the Poplars site.

Another record-breaking machine to work at Poplars was the largest wheel dozer seen in the UK, the Michigan 480 imported from America. Powered by a 600-horsepower Cummins V12 diesel, it weighed 53 tons equipped with dozer blade and push plate for push-loading scrapers in record time.

Robert McGregor & Sons Ltd.

Robert McGregor & Sons Ltd. was another civil engineering contractor to enter the opencast scene. Headquartered in Greater Manchester, it ran a local office and equipment yard near Chesterfield, a central location for the Yorkshire and Derbyshire coal fields. This company depot encompassed a maintenance shop capable of performing major equipment rebuilds and modifying machines to suit McGregor's special requirements.

The company enjoyed a long run in opencast mining, becoming Norwest Holst Mining in 1981. At its peak it ran a sizeable fleet of equipment including Lima 2400s and two Ruston-Bucyrus 110-RB diesel electric draglines. One of these, along with a seven-yard Lima 2400 and a 2½-yard Ruston-Bucyrus 54-RB shovel, was employed on the Blue Lodge site near Stavely, Derbyshire. In order to accommodate the needs of this compact but deep site, McGregor placed the 54-RB shovel in the bottom of the pit, excavating the 'key cut' against the high wall to assist the seven-yard Lima dragline. The shovel excavated the tough material, dumping it within reach of the dragline; greatly improving the latter's production. Blasting was not allowed because of proximity to local residents.

The 110-RB dragline was mainly employed on the spoil side of the cut, pulling back the material dumped by the 2400 to gain additional spoil room. Some of this material would thus be handled three times, once by each of the three machines, a necessary procedure given the machines' limited capabilities on such a restricted site.

On completion of Blue Lodge in 1967, McGregor moved its equipment to a follow-on site at Street Lane, Denby, Derbyshire. The diesel-electric 110-RB dragline, with 100 feet of boom and carrying a 4½-yard bucket, received an overhaul and coat of paint showing a new McGregor logo. McGregor also employed a fleet of Euclid TS-14 scrapers push-loaded by a 162-horsepower Michigan 180 wheel dozer. The faster wheels, as opposed to the usual crawler tracks, increased scraper production in dry weather, but it was a different story when rain clouds appeared, as tyres easily lost traction.

McGregor progressed through the 1960s and its experience prepared it well for one of the company's largest opencast contracts, the Shilo site near Cotmanhay, Derbyshire, obtained in 1970. Newer and larger equipment was employed at this site, which is described in Chapter 4. McGregor is credited with producing a total of 4.56 million tons of opencast coal.

Northern Strip Mining Ltd.

Northern Strip Mining Ltd. (NSM) of Sheffield, although carrying out general civil engineering and excavation work following forced government cutbacks of coal production in the 1960s, never abandoned its core business of coal-mining. It operated the Penn Hill site at Cudworth near Barnsley, Yorkshire, for about ten years, maintaining coal sales to its customers. This site was initially rejected by the OE because it was considered extensively worked-out by old underground mining. But after taking over a licence for the site, performing its own exploration drilling, and developing a mine plan, NSM found a high quality, solid five-foot thick coal seam! A two-yard Lima shovel, Ruston-Bucyrus 43-RB 1¾-yard dragline and Smith 21 draglines were the main stripping machines here. A few miles to the north, NSM operated another Lima 802 shovel at its Bircroft site at Sharleston near Wakefield. These were private sites, not operated under contract to the OE.

In the mid-1960s, NSM was also successful in obtaining a long-term but relatively small opencast coal site at Brierly Hill near Stourbridge, Worcestershire. This site turned out to be very profitable and financially supported the company during the lean years of the 1960s. The site was scattered with abandoned industrial dereliction, including old buildings, waste material, toxic ponds and surface remains of old underground mines including dangerous mine shafts. Collapsed workings had caused subsidence that delayed new development, but following site rehabilitation, subsidence from underground workings was eliminated. Backfilled material was compacted and the surface drained and re-contoured in preparation for property development. Today Withymoor Village occupies the site, with its 2,000 houses, schools, shopping centre, public house, modern industrial development and green space with a lake.

By the end of the 1960s, NSM began tendering for OE contracts for much larger sites, as opencast coal production rapidly rose again in the 1970s. The company pioneered large hydraulic excavators in the UK, purchasing the very first O&K models RH60 (nine cubic yards) and RH300 (30 cubic yards), and became one of the largest UK coal companies.

Murphy Brothers Ltd.

Murphy Brothers Ltd. of Thurmaston, Leicester, was another prominent earthmoving contractor that also owned a transport division with a large fleet of road lorries. It operated several large private sites during the 1960s, and also obtained some coal contracts from the OE. Mostly concentrated in central England, private sites were located at Lower Gornal near Dudley, Overseal near Burton-on-Trent and Brierly Hill near Stourbridge. The latter was adjacent to the NSM site just described, where similar activity took place to remove coal and transform industrial dereliction. A large fleet of different excavating machines was put to work on this site, notable for its restricted working area and close proximity to residential property. Overburden stripping machines in use included a 4½-yard Marion 111-M dragline, a seven-yard Lima 2400 dragline and a 4½-yard diesel-electric 110-RB shovel.

One of Murphy's biggest sites was the Newman Spinney Restoration project near Killamarsh, Derbyshire. The project resulted from a failed experiment carried out jointly by the Central Electricity Generating Board and the East Midlands Electricity Board to gasify coal underground by applying heat. Fires developed, and the only firm solution was to excavate the entire affected area and remove the remaining coal to put out the fires. Work started in 1962 and involved excavating some 26 million cubic yards of overburden and retrieval of any remaining coal. The Newman Spinney site is described in more detail in Chapter 7. Murphy Brothers' coal operations expanded rapidly in the 1970s with contracts obtained from the OE extending from Scotland to South Wales.

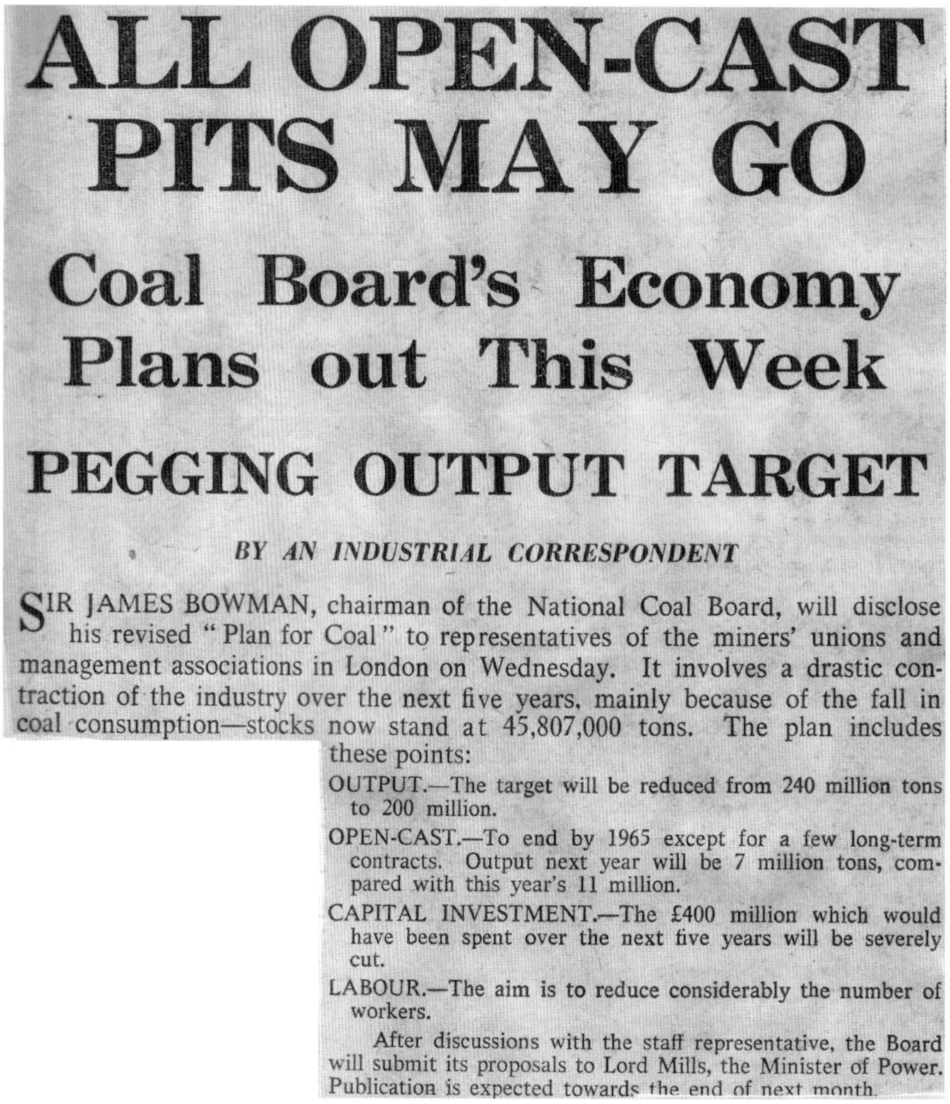

Figure 3.1 NCB chairman Sir James Bowman announces drastic contraction of coal industry in 1960 and predicts all opencast to go by 1965.

ALL OPEN-CAST PITS MAY GO

Coal Board's Economy Plans out This Week

PEGGING OUTPUT TARGET

BY AN INDUSTRIAL CORRESPONDENT

SIR JAMES BOWMAN, chairman of the National Coal Board, will disclose his revised "Plan for Coal" to representatives of the miners' unions and management associations in London on Wednesday. It involves a drastic contraction of the industry over the next five years, mainly because of the fall in coal consumption—stocks now stand at 45,807,000 tons. The plan includes these points:

OUTPUT.—The target will be reduced from 240 million tons to 200 million.

OPEN-CAST.—To end by 1965 except for a few long-term contracts. Output next year will be 7 million tons, compared with this year's 11 million.

CAPITAL INVESTMENT.—The £400 million which would have been spent over the next five years will be severely cut.

LABOUR.—The aim is to reduce considerably the number of workers.

After discussions with the staff representative, the Board will submit its proposals to Lord Mills, the Minister of Power. Publication is expected towards the end of next month.

Figure 3.2 A fleet of Euclid R-22 dump trucks line up at the end of shift at Wimpey's Dark Lane opencast site at Adlington, Lancashire in 1960. For the first three decades of British opencast coal, G. Wimpey & Co. Ltd. was the largest producer, achieving more than 37 million tons by 1974. *KH*

Figure 3.3 At the long-running Dunraven site near the village of Rhigos above the Vale of Neath, South Wales, Parkinson employed this Bucyrus-Erie 190-B shovel, seen here loading a Euclid R45 dump truck. A few of these nine-yard shovels were imported from America until 1974 when Ruston-Bucyrus started manufacturing the 12-yard 195-B at Lincoln.

Figure 3.4 One of Parkinson's Euclid R-65 haul trucks photographed at Dunraven in 1969. First built in Scotland in 1968, the 65-ton capacity hauler was the largest yet built in the UK. The model was upgraded to the R-70 in 1972. *KH*

Figure 3.5 Photographed in April 1969, these Euclid R-65 haul trucks are en route to Parkinson's Dunraven site. With an escort van, they wait for passing motorists on the narrow A1 road at Thrimby, just south of Penrith where the nearby M6 motorway was still under construction. The 65-ton trucks were 'roaded' the entire distance from the factory in Scotland to South Wales. *KH*

Figure 3.6 An aerial shot of Parkinson's Dunraven site in 1969. The site is well advanced, having received several extensions. As well as numerous coal excavators and haul trucks, three main excavators can be seen removing overburden. Nearest is a Ruston-Bucyrus 110-RB dragline with 4½-yard bucket, while Bucyrus-Erie 150-B and 190-B shovels with six- and nine-cubic yard dippers respectively are seen in the middle and rear distance.

Figure 3.7 One of two Rapier W300 walking draglines purchased new by Parkinson for work at Heddon-on-the-Wall, Northumberland. This picture was taken at the Bays Leap site in 1965. Its mechanical and electrical design included Ward-Leonard control and single motor for hoist and drag. *KH*

Figure 3.8 Working on the spoil side this seven-yard Rapier W300 dragline pulls back spoil to gain more room at Bays Leap site, Northumberland. The W300 weighed some 420 tons, considerably more than the Lima 2400 of similar capacity, but with a 140-foot boom, its reach was much greater. *KH*

Figure 3.9 The stalwart Lima 2400 was the shovel of choice for rock and overburden removal on British opencast sites throughout the 1960s and the early 1970s. It operated as a seven-yard shovel or with dragline buckets up to eight cubic yards. Its diesel power gave it superior advantage for short-lived sites typically found in the UK where the expense of electrification was not justified. This 2400 loads an Aveling-Barford SN 35-ton dump truck for Derek Crouch (Contractors) Ltd. at its Lumley Castle site, Durham.

Figure 3.10 Activities at Crouch's Lumley Castle site show a Ruston-Bucyrus 22-RB and NCK 605 shovels at pit floor level excavating and loading coal. Notice the coal lorry has been 'side-boarded' to gain extra volume, allowing its rated capacity to be reached with the low density coal. This practice highlights the skill of the shovel drivers, as one swipe of the swinging open dipper door could destroy the crafted wooden structure.

Figure 3.11 Activity at Shortwood Opencast Coal Site, Trowell, Derbyshire, in June 1964. Coal in the foreground is loaded by a Ruston-Bucyrus 22-RB shovel into a Bedford steel-bodied lorry. The Caterpillar D4 bulldozer prepares a smooth road for the lorry, not designed for off-road use, to leave the rough site area and haul to a disposal point a few miles from the site. Two Ruston-Bucyrus 54-RB draglines, with 2½-cubic yard buckets, are seen in the background removing overburden from this relatively shallow seam. *KH*

Figure 3.12 W.J. Simms Sons & Cooke Ltd. employed a variety of tractors and scrapers at Shortwood opencast site. In this picture some are removing overburden, but they were also used to salvage subsoil and topsoil prior to excavation by draglines. The soil would be stockpiled or placed on land reclaimed after mining. Shown are two LeTourneau Model D Tournapull motor scrapers of seven cubic yards capacity, assisted by a Caterpillar D7 as a push tractor. Behind these, a Caterpillar D8 tractor pulls a LeTourneau scraper that is about to be pushed by a second D8. Note the makeshift canvas cab on the D7 offering almost zero visibility for the driver. *KH*

Figure 3.13 More scraper action at Shortwood with a Caterpillar D8 and LeTourneau 12-yard scraper in the background. A second D8 is aligning itself to push the scraper during loading. In front, a D7 tractor is pulls a cable-operated ripper made by Ruston-Bucyrus. Extra push and down pressure is being applied by a D8 fitted with hydraulically-operated blade. *KH*

CHAPTER 3   THE 1960s   **73**

Figure 3.14 At Simms' Shortwood site, topsoil and subsoil are salvaged by a D8 tractor pulling a 'modern' Caterpillar low bowl scraper with pushing assistance provided by an identical D8. After removal of brush and timber (notice the pile stacked at right ready for burning), topsoil and subsoil were stripped and stored in separate piles for later use. After coal was removed and areas were backfilled and contoured, the same scraper fleet would place soil material in specified layers in the restoration process. *KH*

Figure 3.15 This Ruston-Bucyrus 110-RB diesel-electric dragline was the largest excavator at Shortwood. It was hired by Simms from opencast contractor Shephard-Hill & Co. to handle some tough rocky overburden. As no blasting was allowed in close proximity to local residences, huge pieces of rock gave the 4½-yard dragline a tough workout. A 100-foot long boom was standard for this bucket size, and the big V12 Ruston-Paxman diesel engine driving the motor-generator set put out 528 horsepower. *KH*

74 CHAPTER 3 BRITISH OPENCAST COAL

Figure 3.16 A number of coal seams being mined at Shortwood. On the top level a LeTourneau D Tournapull removes subsoil. Below that, a Caterpillar D7 and Onions scraper clean off the upper coal seam in preparation for loading. The next seam down is being loaded by a Ruston-Bucyrus 19-RB shovel into a Ford Thames Trader lorry. The exposed high wall in the foreground will lead to further coal seams at lower levels. *KH*

Figure 3.17 Two Ruston-Bucyrus 54-RB draglines work at Simms' Doe Lea opencast site adjacent to the M1 motorway, Junction 29 near Chesterfield in 1968. Overlooked by Hardwick Hall, this site uncovered several shallow seams with the 2½-yard draglines. The five exhaust pipes on the 54-RB in the foreground indicate power comes from a five-cylinder Ruston VCBN diesel engine running at a low 450rpm, typically fitted to 54-RBs up to the early 1950s. The later 54-RBs were fitted with the V6-cylinder Ruston-Paxman 6RPHN, which ran at 950rpm. *KH*

Figure 2.18 Doe Lea was one of several short-term opencast sites worked in Derbyshire to prepare for the M1 motorway construction. On these sites, backfilled material placed on the alignment of the motorway was compacted according to Ministry of Transport specifications. This Lima 2400 six-yard shovel loads a 35-ton Aveling-Barford SN-35 haul truck, while a Caterpillar D8 dozer handles clean-up. As on other sites, Simms made use of surplus machines from the extensive fleet of Shephard-Hill & Co., no doubt under suitable hire arrangements. *KH*

Figure 3.19 Another site in the path of the M1 motorway, successfully executed by Simms, was Sunnyside near Tibshelf, Derbyshire. The main excavator here is a Ruston-Bucyrus 54-RB 2½-yard dragline, helped by a fleet of Euclid TS-14 motor scrapers and several of Simms' old Caterpillar D8 crawler tractors, some pulling LeTourneau scrapers and others equipped with dozer blades. This picture, taken in 1966, shows Sunnyside in its final stages of completion. The 54-RB dragline has finished its work and the fleet of dozers and TS-14 scrapers is completing the backfill. The D8 tractor pulls a large deadweight roller to compact the fill in readiness for the M1 motorway. The village of Tibshelf is in the background. *KH*

Figure 3.20 A general view of the Sud site at Barlow near Chesterfield. Two Lima 2400 shovels remove overburden, which is being drilled in preparation for blasting by a mobile Reichdrill. Beginning in 1958, more than one million tons of coal were extracted during the course of this site by contractor Martin Cowley Ltd. *KH*

Figure 3.21 A Lima 2400 excavator at Martin Cowley's Sud site gives a 22-ton Euclid R-22 a full load from its six-cubic yard dipper. One of the most popular medium-sized excavators, the diesel-powered 2400 enjoyed virtually no competition in its size range during the 1960s in the UK. *KH*

CHAPTER 3   THE 1960s   **77**

Figure 3.22 Grossly overloaded, this 22-ton Euclid arrives at the dump with an enormous boulder, but its hydraulic hoist was unable to lift the body. Assistance had to be applied from another truck before the load could be tipped out. *KH*

Figure 3.23 Three Lima 2400 six-yard shovels handle the bulk of overburden excavation at Sud site. Tough rocky material and limited blasting challenged men and machines. *KH*

Figure 3.24 One of Cowley's new Euclid R-27 trucks deposits its 27-ton load over the edge of the dump at the Sud site. The R-27 became available from the Euclid (GB) factory from 1957. *KH*

Figure 3.25 One of the Lima 2400s at Sud Site was frequently changed from shovel to dragline. Here the dragline removes the lower interburden to uncover the lowest coal seam. *KH*

Figure 3.26 A rare British-built Blaw-Knox BK-12 grader assists with haul road maintenance at Sud site. Weighing 24,000 pounds, this tandem-drive machine competed with another British grader, the Aveling-Barford 99-H of similar operating weight. Both models were powered by the same 120-horsepower Leyland engine. *KH*

Figure 3.27 In the final restoration at the Sud site, two of the Lima 2400 shovels were moved to the top of the dump to begin the major task of backfilling the final void. In 1965, Cowley purchased a new fleet of Euclid R-45 trucks for this stage. *KH*

80 CHAPTER 3 BRITISH OPENCAST COAL

Figure 3.28 Two of Cowley's new R-45 Euclid haulers dump their loads under the care of the 'tip man' at the Sud site in 1965. *KH*

Figure 3.29 A Ruston-Bucyrus 38-RB takes on some heavy digging at Sir John Jackson's Carway House opencast coal site near Trimsaran, South Wales. Two Lima 2400 draglines in the background remove upper layers of overburden.

Figure 3.30 One of three Lima 2400s at work on Jackson's Carway House opencast site near Trimsaran, South Wales. Standard arrangement for the 2400 dragline was seven cubic yards on a 120-foot boom.

Figure 3.31 A Ruston-Bucyrus 38-RB dragline removes overburden at one of Jackson's sites in South Wales, while a team of Caterpillar D7 crawler tractors pulling Onions scrapers remove the last layers of overburden above the coal seam.

Figure 3.32 One of the longest-running sites in central England was Poplars near Cannock, Staffordshire, operated by R.A. Davis (Midlands) Ltd. A Ruston-Bucyrus 22-RB loads coal, and in the background one of R.A. Davis' Lima 2400s loads an Aveling-Barford SN 35-ton dump truck.

Figure 3.33 A pair of brand new Ruston-Bucyrus 22-RB shovels work for R.A. Davis at the Poplars opencast site in 1964. Beginning in the mid-1950s, the site boasted a diverse fleet of equipment including many on test by their manufacturers or sales agents. Some of the larger machines included five crawler draglines: four Lima 2400s and a Marion 111-M.

Figure 3.34 The Blue Lodge site was operated by Robert McGregor & Sons Ltd. in the mid-1960s. The 110-RB dragline pulls back spoil dumped by the Lima 2400 dragline to gain additional spoil room on this deep and compact site. At extreme right, a Ruston-Bucyrus 22-RB loads coal. *KH*

84 CHAPTER 3 BRITISH OPENCAST COAL

Figure 3.35 Fuelling time for the big 12-cyinder Ruston-Paxman 12 RPHN.II engine in the back of McGregor's Ruston-Bucyrus 110-RB diesel electric dragline at Blue Lodge. The 528 horsepower engine drives DC generator sets for the Ward-Leonard electric system consisting of independent DC reversible motors for the hoist, drag and swing motions. A Lima 2400 diesel dragline behind the 110-RB also waits for diesel fuel. *KH*

Figure 3.36 McGregor placed a Ruston-Bucyrus 54-RB shovel in the bottom of the pit to excavate the 'key cut' against the high wall to greatly improve the seven-yard Lima dragline's production at the Blue Lodge site. Since blasting was not allowed here because of proximity to local residents, the shovel excavates the tough material and dumps it within reach of the dragline. *KH*

Figure 3.37 On completion of Blue Lodge site in 1967, McGregor moved its diesel-electric 110-RB dragline and other equipment to a follow-on site at Street Lane, Denby, Derbyshire. The dragline received an overhaul and coat of paint showing a new McGregor logo. This machine works with 100 feet of boom carrying a 4½-yard bucket. *KH*

Figure 3.38 At the Street Lane site, McGregor employs a Michigan 180 wheel dozer as a push tractor for a fleet of Euclid TS-14 motor scrapers, greatly increasing their productivity in dry weather. Its 162 horsepower adds to the scraper's 296 horsepower, shortening loading time considerably. The dozer then reverses back at high speed to meet the next scraper arriving in the cut, increasing overall scraper production. *KH*

86 CHAPTER 3 BRITISH OPENCAST COAL

Figure 3.39 Northern Strip Mining Ltd. of Sheffield (NSM) operated the Penn Hill site at Cudworth, Barnsley for about a ten-year period starting in the mid-1950s. Here a 1¾-yard Ruston-Bucyrus 43-RB dragline uncovers the high-quality, solid five-foot seam that made this site very profitable. *KH*

Figure 3.40 General foreman Barney Kelly stands by a ⅜-yard Ruston-Bucyrus 10-RB coaling shovel at NSM's Penn Hill site. *KH*

Figure 3.41 This photo, taken in 1962, shows a ¾-yard Smith 21 shovel, built at Rodley near Leeds, loading a British-built Euclid R-15 haul truck. The thick five-foot seam was a good match for this shovel, regarded at the time as quite a large machine to load coal on a private site. *KH*

Figure 3.42 This American-built Lima 802 two cubic yard shovel, one of the main stripping machines at Penn Hill, worked with 15-ton Euclid R-15 haul trucks. The Ruston-Bucyrus 43-RB dragline can be seen in the background. *KH*

88 CHAPTER 3 BRITISH OPENCAST COAL

Figure 3.43 In the lean opencast years of the early 1960s, Northern Strip Mining Ltd. kept its machines busy by undertaking several industrial development projects. Here a Smith 21 with backhoe equipment loads a Morris 5-ton lorry in Sheffield.

Figure 3.44 A few miles north of Penn Hill site, NSM operated another private site called Burcroft, located at Sharleston, near Wakefield. Running for many years, the comparatively small site employed a two-yard Lima 802 shovel as the main stripping machine. In this photograph, taken in 1961, a Caterpillar 3T-series D7 tractor equipped with LeTourneau cable-operated dozer blade, keeps the tip tidy, as overburden material is delivered by Euclid R-15 haul trucks of 15 tons capacity. *KH*

Figure 3.45 In the mid-1960s, NSM worked the Brierly Hill site near Stourbridge, Worcestershire. This 1966 photograph shows the multiple coal seam operation, with two shovels – Ruston-Bucyrus 54-RB at left and Lima 802 at right – loading overburden into Euclid R-15 and R-24 haul trucks. Caterpillar 631 scrapers and a Euclid C-6 bulldozer clean off the upper benches in preparation for loading the coal. *KH*

Figure 3.46 One of NSM's 2½-yard Ruston-Bucyrus 54-RB shovel at Brierly Hill site. This shovel and a fleet of Euclid R-24 haul trucks were purchased new for the job. *KH*

Figure 3.47 Shown at Brierly Hill site, this 1½-yard Ruston-Bucyrus 38-RB shovel is a good match for the Euclid R-15. The ten-yard R-15 is rated at 15 tons capacity. *KH*

Figure 3.48 Two Ruston-Bucyrus 54-RB shovels and a 38-RB remove overburden at the Brierly Hill site in 1969. Originally scattered with abandoned industrial dereliction, the site when finished was transformed into valuable building property including housing and modern industrial development. *KH*

Figure 3.49 Easy does it! A 54-RB shovel coaxes an oversize rock into a Euclid R-24 hauler at NSM's Brierly Hill site. Again, blasting was not possible here as it was surrounded by residential property. On completion, the site was developed into Withymoor Village, a modern housing project including schools, a shopping centre and a landscaped park with a lake. *KH*

Figure 3.50 Intense activity on Murphy Brothers' Brierly Hill opencast site near Stourbridge, Worcestershire, located adjacent to NSM's site just described. At upper left, a 4½-yard Marion 111-M dragline pulls back spoil deposited by the seven-yard Lima 2400 dragline. The Lima works ahead of a 4½-yard diesel electric shovel, which is loading Euclid R-24 haul trucks. Further to the right, a Reichdrill puts down holes for blasting. Above this a Ruston-Bucyrus 22-RB shovel loads what appears to be old Leyland and AEC lorries and, in the foreground, a 38-RB shovel loads a fleet of Foden 15-ton dump trucks. *KH*

92 CHAPTER 3 BRITISH OPENCAST COAL

Figure 3.51 In the 1960s, Murphy Brothers executed several sites around Overseal in Leicestershire. Coal and fire clay were recovered. A pair of Ruston-Bucyrus 22-RB shovels load the coal, while, at the right, a 38-RB shovel removes rocky parting. A four-yard NCK 1405 shovel in the background loads a 35-ton Aveling-Barford SN dump truck. *KH*

Figure 3.52 This rare NCK 1405 shovel worked at Murphy Brothers' sites at Overseal. Built at Sheffield, the four-yard machine loads an Aveling-Barford SN 35-ton dump truck in about six passes. The SN truck was offered with a choice of diesel engines: a 476-horsepower GM, or 450-horsepower Rolls-Royce. *KH*

Figure 3.53 Murphy Brothers operated several large coal sites in central England including those at Lower Gornal, Overseal and Brierly Hill. This diesel-electric Ruston-Bucyrus 110-RB deposits a 4½-yard load into a Euclid R-24 truck. This type of shovel, free from a trailing power cable, was useful on smaller sites where the cost of electrification was unjustified.

Figure 3.54 Throughout the 1960s, Lomount Construction Ltd. was active in opencast mining in the north-east of England. At its Horsley site, just west of Newcastle, the company mined almost two million tons of coal beginning in 1961. This interesting shot shows a Ruston-Bucyrus 150-RB dragline doing double duty. First it removes a ramp to expose coal below, throwing waste material to the right. Then it removes the coal, dumping into a hopper that will feed coal into road lorries when they arrive. An Aveling-Barford 99-H grader maintains the roads.

94 CHAPTER 3 BRITISH OPENCAST COAL

Figure 3.55 This view, taken at Horsley during the restoration stage in 1965, shows two Ruston-Bucyrus 150-RB shovels, an NCK 605 shovel and a Ruston-Bucyrus 5-W walking dragline. One of the 150-RBs has already moved to the upper spoil piles to begin filling in the final void. The NCK 605 removes the last remaining coal. *KH*

Figure 3.56 In the 1960s, Currall, Lewis & Martin Ltd. ran several opencast site scattered throughout England and Wales. This view shows the company's Yeckhouse site at Consett, Durham, in 1969. A six-yard Lima 2400 fills a Euclid R-45 with a full 45-ton load, while a Ruston-Bucyrus 22-RB clears debris from coal in preparation for loading. A large pile of soil has been salvaged in the background.

Figure 3.57 Sir Alfred McAlpine was another national contractor heavily involved in opencast coal, especially in the early years. But the company withdrew from this work early in the 1960s, a victim of the government's cutback in opencast production. In 1962 its last site was completed, with total coal production of 9.24 million tons achieved from 93 sites since 1943. This view shows early work on an Alfred McAlpine site with a Ruston-Bucyrus electric shovel loading Euclid R-22 trucks, and a 54-RB dragline loading salvageable soil for long haul to a stockpile. In 1987, Alfred McAlpine re-entered the industry when awarded the 300,000-ton St Johns contract in Yorkshire.

Figure 3.58 Coal augering was sometimes used at British opencast sites in certain areas where the surface could not be disturbed. From the base of the high wall, horizontal auger holes extend to more than 100 feet into the exposed seam, and although only 50–60 per cent of the seam is recovered, this method extracts coal that would otherwise be inaccessible.

CHAPTER FOUR

1970-1985

In the previous chapter we saw how British opencast coal production throughout the 1960s was limited to a low average of roughly seven million tons per year. Obtained at a cost well below that from underground methods, it was still enough to financially subsidize underground coal. It also spurred the development of efficient excavation methods and better equipment design technology.

Bulk excavation for construction projects of the 1960s and into the 1970s depended largely on organizational skills learned from opencast experience. Efficient earthmoving machines now became the contractor's tool to complete large earthmoving projects at lowest cost per cubic yard. Fortunately there were many projects to keep contractors busy. The motorway programme was in full swing, and the construction boom of the 1960s and 1970s included several nuclear and coal-fired power stations. Britain's 'New Towns' were also built and many major industrial sites, including several steelworks were completed. All these resulted in earthmoving at unprecedented rates, thanks to the most modern equipment.

Opencast mining, although at a much lower level in the 1960s than the previous decade, was able to vastly improve its restoration and environmental standards. The old adage 'opencast destroys the land' could no longer be applied.

But despite the many positives in favour of opencast coal, the government dealt another blow to the industry in 1967 when it introduced a White Paper on Fuel Policy, heralding a return to the 1960 stance of completely phasing out opencast coal. Contending with a background of overall lower coal demand, the White Paper listed even more restrictive conditions against opencast in an effort to minimize job losses and also to appease a fledgling environmental movement.

In reference to opencast coal, the 1967 White Paper stated: 'The Coal Board considers that, with the use of bigger machines, there should be scope for reductions in cost and that opencast coal should continue to be produced much more cheaply than coal from deep mines. This method of working can, however, cause temporary damage to local amenities. Through screening opencast sites by the erection of embankments and the planting of semi-mature trees, the disturbance has been reduced and the Board is working closely with local authorities.' It also stated: 'The case for continued opencast working, taken by itself, is quite strong, but it has to be related to circumstances of the industry as a whole … and on the level of employment in deep mines.'

After further lengthy wording the White Paper concluded with the following paragraph: 'Production of coal at opencast sites employs comparatively few men for each ton of coal won, and reducing it therefore gives rise to fewer manpower difficulties than reducing deep-mine production. Though opencast is profitable and relatively small, there is no advantage to be gained from continuing it where this can be avoided. The Government has therefore decided not to give further authorization for opencast production except in special cases.'

This alarming White Paper caused the Opencast Executive to reorganise and consolidate several of its regions. But this 1968 reorganization had barely taken effect when it became obvious to all, especially the government, that opencast mining, far from being allowed to run down, needed to expand. Thus this latest directive against opencasting turned out to be even less effective than the earlier one proposed in 1960. The 1967 White Paper on fuel policy turned out to be yet another example of government predictions gone wrong.

Rapid expansion in the 1970s

By 1970, coal supply was already becoming scarce, as demand for electricity from coal-fired generating stations was increasing. The 1972 coal strike by underground miners aggravated the situation and dealt a further blow to the industry, but it did boost opencast coal, mined of course by private contractors with no connection to the miners' unions. Also it was evident that opencast coal stored during the coal strike for periods of up to two years could then be put into normal distribution channels more easily than deep-mined coal. Finally, the insecurity of imported oil and higher prices due to the Arab oil embargo of the early 1970s caused worldwide panic for secure energy, resulting in a massive demand for coal.

With a pending world shortage of oil, and indeed electrical energy from all sources, oil companies began investing in coal by purchasing coal leases or existing mines. Increased coal production was most evident in America where demand for surface mining equipment reached an all-time high, with delivery periods for some of the larger machines quoted at more than four years! In the UK, the government was forced to totally rethink its energy policy, resulting in a complete reversal of the plan to phase out opencast coal. Thus the 1974 'Plan for Coal' now projected a gradual increase of coal production on all fronts, with opencast output to increase to more than 16 million tons per year by the mid-1980s. This output was to be maintained indefinitely.

The 1974 'Plan for Coal' was great news for opencast coal contractors. Finally released from its shackles, the industry expanded at a feverish pitch not seen since the first years of opencasting in the 1940s. The Opencast Executive (OE) of the National Coal Board (NCB) had a large inventory of potential sites poised for this opportunity, and they were tendered to contractors at a brisk pace.

The largest opencast contractors – Wimpey, Parkinson, Crouch, Costain, Taylor Woodrow, Shand and Miller – took immediate advantage of this valuable opportunity to tender for the largest sites and re-equipped their earthmoving fleets with modern machinery. Smaller contractors were by no means left out: some returned to the industry after exiting during the lean years; others who had managed to survive with lower production levels in the 1960s now anticipated expansion and modernized their plant fleets; newcomers joined the industry.

The most notable newcomer was A.F. Budge (Contractors) Ltd., which obtained its first opencast coal contract from the OE in 1974: Biggin Lane in County Durham. By 1978 the company was operating six sites and, after a decade of meteoric growth, boasted some 14 million tons of coal under contract with the OE, the third largest of any UK contractor at that time.

With the 1974 'Plan for Coal' gaining momentum, the OE embarked on a seven-year £33 million purchasing programme during which it purchased 31 major items of earthmoving equipment including large walking draglines, smaller diesel crawler draglines, electric and diesel mining shovels, and 127 dump trucks of various capacities. Because of high world demand for opencast equipment at that time, the OE would order machines well in advance of need and begin financial arrangements for purchase. As contracts were awarded, a machine would be allocated to a site and financial arrangements, through lease or purchase, would be transferred 'at cost' to the contractor. In certain cases, especially with larger shovels and walking draglines, the OE retained ownership.

Often the OE would order machines before deciding where they would be allocated, or which contractor would operate them. Smaller machines, such as five-yard Bucyrus-Erie 88-B shovels and other equipment, would be similarly ordered for almost immediate availability upon award of a contract. The NCB ran a storage and maintenance facility at Wentworth, Yorkshire, where complete machines or parts of machines were stored until needed. There would be an inventory cost to this method, but it did ensure the right equipment was available without delay once contracts were awarded.

Equipment development

Chapter 5 will describe how the OE's fleet of walking draglines was expanded and modernised to accommodate larger sites, and Chapter 6 will examine the impact of hydraulic excavators, a new type of machine that would revolutionise the British opencast industry. Other types of machines also evolved in the 1970s.

Crawler draglines

A wider variety of crawler draglines appeared, beginning with the seven-yard capacity Manitowoc 4600 in 1971. This machine competed with the popular Lima 2400, which for years had dominated the market in the medium-size class. The first was operated by Ruddock & Meighan at its Nackshiven site in County Durham. Imported into the UK by dealer A. Long & Company, the 4600 was not as popular in the UK as it was in the USA, where it gained an excellent reputation for high reliability and where many still operate today. Nevertheless, some 14 4600s were reportedly brought into the UK to work on opencast sites. Contractors, led by Wimpey in 1973, ventured into larger crawler draglines, the first of which was the Marion 195-M at 14 cubic yards capacity. This machine proved so successful that Wimpey purchased another four; Lomount purchased another for its Tanners Hall site in 1980.

Not so successful was the larger Manitowoc 6400 dragline, brought in by the OE to work at Shephard Hill's Low Close site in Cumbria (1975–1981). The 15-yard machine was pre-ordered before being fully tested in America, and suffered some teething problems with its giant swing bearing. Although the problems were remedied by Manitowoc, the machine was rebuilt and sold back to America and no more appeared in the UK. Another unusual dragline to operate in the UK was the German-built Weserhutte SW530 in the seven-yard class, put to work at the Kiersbeath site in 1982 by French-Kier Mining.

Rope shovels

Rope shovels also increased in size. With a world demand for surface mining of all types, Ruston-Bucyrus commenced manufacture of the 195-B, pushing its largest British-built shovel up to the 12-yard class. A total of 44 of these electric shovels were built from 1974 to 1983. Of these, 29 were placed in service on UK opencast coal sites, the first ordered by Wimpey for its Auchingilsie site in Scotland.

Such was the demand for this size of shovel, and because supply was at a premium, the OE ordered five P&H 1900AL shovels from Harnischfeger Corporation in America. The first of these 11-yard shovels was installed at Wimpey's Maesgwyn site for testing in 1976. Others were allocated to sites operated by Wimpey, Fairclough-Parkinson, Taylor Woodrow and Shephard Hill. The 1900AL shovels incorporated the patented P&H Magnetorque drive system, which controlled the rotation of the hoist drum through a variable magnetic field between the drum and an AC motor. DC motors powered the crowd, swing and propel motions.

Dump trucks

Dump trucks also experienced major transformation during this heyday of British opencast coal in unison with changes in loading equipment, all with the objective of increased efficiency. Average dump truck size moved up from the common 45-ton in the 1960s to the 85–100-ton sizes by 1990. In 1967, Shand pushed the threshold upward with its 75-ton Wabco 75A trucks at Glyn Glas, followed by Crouch in 1970 with its 100-ton Unit Rig M85s at Llanilid. In 1981, Northern Strip Mining introduced 170-ton Wabco 170Cs at its Godkin site, followed by the 195-ton Caterpillar 789s operated by Budge at West Chevington in the mid-1980s.

Size increases

Evidence of increasing size of excavators is found in Table 4.1, which indicates 90 excavators having a capacity of more than eight cubic yards working on UK opencast coal sites in 1982. Compare this with the total of 14 shown in Table 2.1. Even greater numbers were achieved with the introduction of the most popular large excavator, the O&K RH-120C in 1983. Some 96 of this 15-yard model alone were imported in succeeding years.

TABLE 4.1 Excavators greater than eight cubic yards capacity working in the UK in 1982

Make	Model	Size (cubic yards)	Number at work 1982
Bucyrus-Erie	1150-B	25	4
Bucyrus-Erie	1550-W	65	1
Demag	H-111	8.5	2
Demag	H-241	18	3
Liebherr	R991	13	1
Manitowoc	6400	15	1
Marion	7400	12	2
Marion	7500	20	1
Marion	7800	30	1
Marion	182-M	10	4
Marion	183-M	9	1
Marion	195-M	14	5
Marion	191-M	15	2
O&K	RH-300	34	1
O&K	RH-75	10	12
P&H	1200	15	5
P&H	1900AL	11	5
Poclain	1000CK	11.5	5
Rapier	W1800	40	1
Rapier	W2000	31	2
Rapier	W700	13	2
Rapier	W600	11	2
Ruston-Bucyrus	1260-W	32	4
Ruston-Bucyrus	195-B	12	19
Ruston-Bucyrus	380-W	12	3
Ruston-Bucyrus	480-W	18	1
		Total	90

The OE's new-found ability to obtain larger sites of longer duration promoted the rapid increase in the use of larger hydraulic excavators, walking draglines and other modern equipment. It allowed contractors to plan sites for optimum efficiency, and records show a remarkable upsurge in productivity. Another advantage of using high-production machines was that smaller sites could often be finished earlier than estimated, thus minimizing any local disturbance.

Significant sites in Yorkshire and Derbyshire

In 1973, James Miller & Partners received the first contract at St Aidans site, adjacent to the River Aire near Woodlesford, Yorkshire, to recover 1.8 million tons of coal. The site started with one of the OE's Bucyrus-Erie 1150-B 25-yard walking draglines, which was moved from the former Poplars site near Cannock and erected at St Aidans in preparation for the new contract. Miller added a Manitowoc 4600 with eight-yard bucket, a six-yard Lima 2400 shovel and an O&K RH60 hydraulic shovel.

In 1981, the site received a major expansion when Miller Mining was awarded the St Aidans Extension contract. For this, the OE provided a new 40-yard Rapier W2000 walking dragline, which joined the 1150-B to work in the same pit. The shovel fleet was also boosted by the addition of three Ruston-Bucyrus 195-B 12-yard electric shovels, an 18-yard Demag H-241 hydraulic backhoe and an older seven-yard Ruston-Bucyrus 150-RB. As expected, a massive haul truck fleet served the shovels, mainly consisting of 85-ton Caterpillar 777s, Euclid R-85s and R-100s.

An interesting feature of the site was that 80 per cent of coal output was hauled to an adjacent loading dock on the Aire and Calder Navigation canal system and shipped by barge to the main customer, Ferrybridge C Power Station. The St Aidans Extension contract called for the extraction of 6.6 million tons of coal with completion set for 1991. But nobody dreamed it would take until 2003 before the site was finished!

In March 1988, operations were brought to an abrupt and untimely halt when the catastrophic failure of an uncharted geological fault under the River Aire caused a major breach in the river bank. For three days, the entire flow of the River Aire flowed into the site, and the resulting flood filled the St Aidans one-mile-long open cut with 22 million cubic yards of water, leaving the deepest part of the site submerged to a depth of 230 feet. Fortunately, Miller was able to remove all personnel and equipment, including the large draglines, safely out of the pit before it was inundated. The prolonged and strenuous efforts, involving several government agencies, to bring the site back into production are beyond the scope of this book, but a brief account of the ten-year saga is described in the Epilogue.

Long-established contractor Shephard, Hill & Co. Ltd. was deeply committed to opencast coal in the 1960s and 1970s. The Steadmill group of sites at Alfreton, Derbyshire operated with an interesting variety of machines. These included two Lima 2400s, one dragline and one shovel, a five-yard Ruston-Bucyrus 110-RB diesel-electric dragline, a new Ruston-Bucyrus 71-RB 3½-yard shovel, a fleet of Caterpillar 627 motor scrapers and associated dozers. A ten-yard Caterpillar 992A wheel loader and a similar-sized Terex 72-81 assisted with overburden removal. Haul trucks included fleets of 35-ton Aveling-Barford SN and Centaur models, and even a couple of 12-ton Scammell Mountaineers. Ruston-Bucyrus 22-RB shovels handled coal loading.

One of the more prominent sites in Derbyshire in the 1970s was Shilo near Eastwood. It started in 1970 and was carried out in three phases by Robert McGregor & Sons Ltd. Two diesel crawler draglines – an eight-yard Lima 2400 and a five-yard Ruston-Bucyrus 110-RB – were moved in from McGregor's former Street Lane site near Denby, Derbyshire. These were supplemented by a brand new American-built five-yard Bucyrus-Erie 88-B shovel, one of six of its type imported in 1970 for use by British opencast contractors. A feature of Shilo was the excavation of a major box cut (initial cut) where the material was transported a significant distance to a storage dump for backfilling a future void. Costly at first, this method pays off at the end of the job when the final void will be quickly filled with a short-haul distance. The final phase, known as Shilo South Extension, commenced in 1978 and was completed in 1981.

Another Derbyshire site starting in 1970 was Shipley Lake near Heanor. Shand Mining operated the site with a variety of equipment including another of the 88-B five-yard shovels. Shand also brought in two eight-yard draglines: a new updated Lima 2400B model and a Manitowoc 4600. The Shipley Lake site involved the removal and reclamation of extensive dereliction left by former industrial activities. These included surface facilities of an old underground mine with pit head gear and mine shafts to be filled. On completion of coal extraction and land reclamation, the site became the present Shipley Lake Country Park and also the site of the now defunct (and levelled) American Adventure theme park. Remains of the old pit head gear with winding pulleys are now preserved as a piece of industrial archaeology.

Significant sites in South Wales

The deep valleys of South Wales are famous for their long tradition of coal-mining. In fact mining in these valleys can be traced back to the very first use of coal. Of course underground mining was the primary method employed in the early days until the early 1940s. At this time, as in other areas of the UK, opencast mining was introduced into South Wales due to the high demand for coal and shortage of manpower during World War II.

The geology in the coal-mining areas of South Wales is extremely complicated. The topography consists of barren mountains interspersed with deep eroded valleys. The coal seams have been dramatically folded into anticlines and synclines, and are severely faulted. In many ways this complicated geology favours opencast mining methods because the entire coal seam can be removed in the open, regardless of its shape or angle. Many opencast sites have been – and still are – extracting coal that would otherwise be impossible to mine by underground methods. The complicated geology and difficult access is further compounded by the climate here, with South Wales experiencing some of the highest rainfall figures in the UK.

But difficult mining conditions have their rewards. The type of coal mined in South Wales is known as anthracite, a rare commodity found in only a few locations in the world. Anthracite boasts the fewest impurities and the highest calorific value of all types of coal. It also has the highest carbon content, ranging between 92 and 98 per cent. It burns so hot that it must be used in special furnaces designed for the purpose. With all these advantages in its favour, anthracite commands a higher selling price than other ranks of coal.

Higher coal prices mean that mining ratios at opencast sites can be correspondingly higher. Ratios of up to 30:1 are not uncommon, i.e., 30 cubic yards of rocky overburden must be removed to release one ton of coal. But efficiency in such conditions is still essential, and this has been demonstrated over the years by the opencast coal contractors who continually strived for top productivity with modern equipment, often some of the largest available in the world.

In 1970, major opencast contractor Derek Crouch made headlines by announcing the OE had awarded it a major new coal site in South Wales. The site, named Llanilid near Llanharan, Glamorgan, would yield seven million tons of anthracite over a period of 15 years. The geologically difficult site with a high coal-to-overburden ratio consisted of multiple seams, mostly steeply dipping and faulted. The site would also reclaim and remove a significant amount of industrial dereliction including surface facilities from old underground mines.

The big news in this announcement, reported in several trade magazines, was that Crouch would import 11 dump trucks from America, the largest ever seen in the UK. They were diesel-electric drive model M85 trucks of 100 tons capacity supplied by Unit Rig & Equipment Company of Tulsa, Oklahoma, from its Canadian factory. Powered by a 1,000-horsepower engine connected to a DC generator, they cost £100,000 each at the time. After erection at Barry docks, the 17-foot eight-inch-wide vehicles were driven in convoy by road to the Llanilid site on a Sunday, with police escort of course.

Big trucks were not the only new machines Crouch purchased for Llanilid. Two new Marion electric shovels – a ten-yard 182-M weighing 356 tons and a 15-yard 191-M tipping the scales at 445 tons – would excavate the tough overburden and be teamed with the new trucks. A Ruston-Bucyrus six-yard 150-RB shovel was also on site. Coaling equipment included Ruston-Bucyrus 30-RB shovels, Caterpillar 966 wheel loaders and Poclain hydraulic backhoes fitted with special high-volume coal buckets. Working from the top, backhoes were particularly useful to cleanly remove thin and steeply dipping seams. The coaling excavators loaded a fleet of 15-ton AEC 6 × 4 trucks with coal bodies. Support equipment on site included Caterpillar D8 and D9 bulldozers.

A small but difficult site was undertaken by Wimpey at Ty-Uchaf near Ammanford. Thin coal seams here were variably inclined up to 60 degrees from the horizontal, and geological faulting complicated the working method. Ruston-Bucyrus 22-RB face shovels were the main coal-loading machines assisted by generous portion of hand labour. This site certainly confirmed that opencast methods were the only way to extract coal in such thin, geologically disturbed seams. A Lima 2400 dragline with 120-foot boom and seven-yard bucket and a Ruston-Bucyrus 150-RB electric shovel were the main overburden excavators.

For many years, major civil engineering contractor Taylor Woodrow operated the Royal Arms group of sites near Dowlais, Glamorgan. High above the deep Welsh valleys, this site mined no less than 13 steeply inclined and faulted coal seams, typical of the region. Taylor Woodrow started excavation in 1958 and employed a variety of equipment over the years, including a Ruston-Bucyrus 150-RB and two 110-RB electric shovels. In 1971, the company purchased a new Marion 183-M dragline, the only one to be imported into the UK. The crawler-mounted machine swung a nine-yard bucket on a 120-foot boom and weighed some 370 tons.

When Taylor Woodrow was awarded the nearby Trecatty opencast site in 1976, the 183-M moved there and was joined by three more electric excavators: a P&H 1900AL and two 12-yard Ruston-Bucyrus 195-B electric shovels. The haulage fleet included four Euclid R-85 dump trucks, a rare type for the UK. The 360-ton 1900AL excavator, one of five imported into the UK, was specified with an 11-yard rock dipper.

The Dunraven site, operated by Parkinson was one of the longest running opencast sites in South Wales, located near the village of Rhigos above the Vale of Neath. Here Parkinson employed its second Marion 7400 dragline with 12-yard bucket, which had been obtained second-hand from Sweden. Other machines included a Marion 191-M with 14-yard dipper, a seven-yard Lima 2400 dragline and Bucyrus-Erie 150-B and 190-B shovels with six and nine cubic yard dippers respectively.

The NCB planned expansion for opencast coal began to materialize in the early 1970s. In fact production forecast to be achieved by the mid-1980s was actually achieved by 1980. With a number of large coal-fired power stations built in Yorkshire during the 1960s committed to long-term coal demand, and a general increase in demand throughout the UK, contractors saw a bright future in coal. The rapid rise in opencast production enabled the cost per ton to decrease even further because of improved efficiency at site level, and the ability to employ bigger, more productive machines on larger sites of longer duration.

Figure 4.1 In 1971, the seven-yard capacity Manitowoc 4600 dragline began to appear on British opencast sites to compete with the Lima 2400. By the end of the decade some 16 of these units were working in the UK. *Manitowoc Engineering*

Figure 4.2 Contractor Wimpey led the way to larger crawler draglines when it purchased a Marion 195-M. Between 1973 and 1976, Wimpey purchased five of these 500-ton machines, which swung 14 to 17 cubic yard buckets on 130-foot booms. *W.R. Peregrine Studio, Llandeilo*

104 **CHAPTER 4** BRITISH OPENCAST COAL

Figure 4.3 Merriman-Meighan employed this Manitowoc 4600 seven-cubic yard capacity crawler dragline at Coalfield Farm site, Leicestershire. Much of its time was spent as a 'pull back' machine on the spoil piles, creating additional room for spoil deposited by the big 1260-W walker. *KH*

Figure 4.4 The biggest crawler dragline to work in the UK was this Manitowoc 6400 erected in 1977 at Shephard Hill's Low Close site in Cumbria. Not entirely reliable, the 15-yard machine is shown undergoing major repairs assisted by a Manitowoc crawler crane. On site completion, the 6400 was sold back to America. *KH*

Figure 4.5 A Ruston-Bucyrus 71-RB 3½-yard shovel loads an Aveling-Barford SN 35-ton dump truck at Shephard Hill's Steadmill site at Alfreton, Derbyshire. Although small by later standards, the two machines were a perfect match for each other. *KH*

Figure 4.6 An Aveling-Barford Centaur hauler is loaded by the Ruston-Bucyrus 71-RB shovel at Steadmill. The British-built Centaur dump truck range, from 35 to 50 tons capacity, was quite popular on opencast sites, with large fleets at Anglers in Yorkshire and Furnace Hillock in Derbyshire. *KH*

Figure 4.7 No place to put the dirt! Sometimes at the end of a cut, excavating box cuts or in tight corners of a site, loading haul trucks by dragline is the most economical solution. Here an Aveling-Barford SN dump truck receives an eight-yard bucketful from a Lima 2400 dragline at Steadmill, Derbyshire. *KH*

Figure 4.8 Steady does it! One bucket from this Lima 2400 gives a full eight-yard load on the Scammell Mountaineer dump truck at Steadmill. *KH*

Figure 4.9 This action shot at Steadmill shows the Lima 2400 6-yard shovel side-casting overburden into the previously mined-out cut. A Ruston-Bucyrus 22-RB excavates coal in the foreground. *KH*

Figure 4.10 An American-built Terex 72-81 assists with overburden removal at Steadmill. The ten-cubic yard wheel loader is in 'dig and carry' mode, dumping the material into the worked-out pit, basically doing the work of a shovel. *KH*

CHAPTER 4 BRITISH OPENCAST COAL

Figure 4.11 Also removing overburden at Steadmill, this first-series Caterpillar 992 wheel loader is similarly rated to the Terex 72-81 at ten cubic yards. It is believed that both machines were on trial for Shephard Hill before a decision was made to purchase. *KH*

Figure 4.12 Lomount Construction operated a number of opencast coal sited in the north-east of England. Here a brand new Ruston-Bucyrus 22-RB shovel loads a nice seam of coal into a lorry for road transport to the nearest disposal point. In 1987, Lomount Construction became part of John Mowlem Construction plc and opencast mining continued under the name Mowlem Mining.

Figure 4.13 Plant owned by Lomount's opencast fleet included this rare Smith 40 shovel and Bucyrus-Erie 88-B dragline. In 1971, Lomount was the first of several contractors to import this five-yard diesel-powered excavator from America, which could be equipped with either shovel or dragline attachments. Smaller than the diesel Lima 2400, it provided its owners with independent flexibility of a substantial machine without the need to electrify a site.

Figure 4.14 Draglines galore at Lomount's Incline site! Located at Stanley Hill near Crook, County Durham, it ran from 1974 to 1979 and produced 684,000 tons of coal. Dragline action (from front to rear) is provided by a Ruston-Bucyrus 110-RB, Lima 2400, Ruston-Bucyrus 5-W and Ruston-Bucyrus 150-RB. Ruston-Bucyrus 22-RB shovels and a Caterpillar wheel loader handle coal from the multiple seams. *KH*

Figure 4.15 In 1970 Robert McGregor & Sons Ltd. commenced its Shilo site in Derbyshire by excavating a major box cut employing three diesel-driven excavators shown in this action shot. The 88-B and 110-RB dragline load a fleet of 45-ton Terex R-45 dump trucks to remove the upper benches at the deeper end, while the Lima 2400 dragline handles overburden at the shallow end. *KH*

Figure 4.16 This five-yard Bucyrus-Erie 88-B, one of six imported in 1970 for use by British opencast coal contractors, works with the Ruston-Bucyrus 110-RB dragline and Terex R-45 trucks to excavate the box cut at Shilo. Soon after the box cut was completed, the draglines continued to mine the site, strip by strip, and the 88-B shovel moved to another job. *KH*

Figure 4.17 Starting in 1970, Shand Mining put to work a new Lima 2400B dragline at its Shipley Lake site near Heanor, Derbyshire. The 'B' series, upgraded to a full eight-yard machine, sported a new operator cab profile. *KH*

Figure 4.18 Shand also operated two of the imported Bucyrus-Erie 88-B shovels at Shipley Lake. This one loads a 35-ton Aveling-Barford SN dump truck. *KH*

Figure 4.19 Ruston-Bucyrus 110-RB and 150-RB electric shovels continued to be used well into the 1980s. This diesel-electric 4½-yard 110-RB earns its keep loading Euclid R-24 trucks for Murphy Brothers in the English Midlands.

Figure 4.20 In the early 1980s there were still more than 25 Ruston-Bucyrus 150-RBs working in opencast, the last one put to work in 1978 by Murphy Brothers at Forth, Lanarkshire. Shand Mining applied this late model version to load Terex R-45 haul trucks.

CHAPTER 4 1970-1985 113

Figure 4.21 In 1974, G. Wimpey & Company Ltd. took over the large Auchingilsie site near Cumnock, Scotland. It became home to the very first British-built Ruston-Bucyrus 195-B shovel, here seen loading a Terex R-50 haul truck built locally at Motherwell, Scotland. The site yielded approximately one million tons of coal.

Figure 4.22 From the 1970s, haul trucks increased in size and sophistication. Here a 90-ton Terex 33-11D hauls overburden at UK Coal's Arkwright site in Derbyshire. Miller Mining obtained an earthmoving contract from UK Coal on this massive site, which lasted more than ten years. *KH*

Figure 4.23 Activities at Fairclough-Parkinson's Anglers site, Yorkshire, in 1976. In the background, one of the company's Rapier W300 draglines is still at work, while hydraulic and cable excavators load coal into lorries for transport to the NCB disposal point off site.

Figure 4.24 Working in the 1970s is the last Rapier W150 walking dragline to work on opencast coal. With a six-yard bucket, it still pulled its weight at Miller's Lodge Hill site in Yorkshire, in production from 1973 to 1976. *KH*

Figure 4.25 This is the 25-yard Bucyrus-Erie 1150-B dragline 'Odd Ball' that worked at the St Aidans site from its start in 1974 until the site was flooded in 1988. Afterwards it was preserved at the site as an interpretive centre by the St Aidans Trust, who took over the machine in 1999. *KH*

Figure 4.26 The main cut at St Aidans in 1974 showing the 1150-B in the background, the O&K RH60 hydraulic shovel, and coaling operations with Hy-Mac and Ruston-Bucyrus shovels. *KH*

Figure 4.27 The third most significant manufacturer of large electric mining shovels was Marion Power Shovel. Not left out of the British opencast scene, the company placed four of its ten-yard 182-Ms with the OE to start the large Butterwell site in 1977 to pre-strip ahead of Big Geordie, the 1550-W dragline. *KH*

Figure 4.28 Crouch operated four of the Ruston-Bucyrus 195-B 12-yard shovels. This is one of two at Acklington, Northumberland, shown filling a Caterpillar 773 50-ton hauler.

Figure 4.29 Another view of Crouch's Acklington site shows the 27-yard Bucyrus-Erie 1150-B dragline at work. Parked for a break is a Bucyrus-Erie 150-B and a fleet of Caterpillar 773 and Unit Rig M-100 haul trucks. Notice the old bus used as a tea shack. The tow bar in front indicates its engine plays no part in its mobility! *KH*

Figure 4.30 In 1976, the OE ordered five P&H 1900AL electric shovels from Harnischfeger Corporation in America. These were allocated to sites operated by Wimpey, Fairclough-Parkinson, Taylor Woodrow and Shephard Hill. They arrived with 11-cubic yard dippers, but at least one was later modified with a 14-yard dipper. This one loads a 50-ton Caterpillar 773 truck for Wimpey at Maesgwyn.

Figure 4.31 In South Wales, Crouch operated two Marion shovels – a 182-M and a 191-M – to remove rocky overburden at the Llanilid site. Here, the 15-yard Marion 191-M, one of the largest in the UK, fills a Unit Rig 100-ton truck. *KH*

Figure 4.32 This pit view of Llanilid clearly shows the complicated geology so typical of South Wales. Working with the thin dipping seams of anthracite, cable and hydraulic excavators and a wheel loader fill 15-ton dump trucks. In the centre, the ten-yard Marion 182-M keeps a fleet of 100-ton Unit-Rig trucks busy. *KH*

CHAPTER 4 1970-1985 119

Figure 4.33 In a single order, Derek Crouch (Contractors) Ltd. ordered 11 of these diesel-electric drive Unit Rig M85 100-ton trucks. Costing £100,000 each in 1970, they were put in service at Llanilid site contracted to recover six million tons of anthracite. *KH*

Figure 4.34 A Poclain backhoe peels off one of the thin coal seams at Llanilid and loads a fleet of AEC 15-ton dump trucks equipped with high-volume coal bodies. *KH*

120 CHAPTER 4 BRITISH OPENCAST COAL

Figure 4.35 One of Wimpey's many Ruston-Bucyrus 150-RB electric shovels excavates some tough rock at the Ty-Uchaf site in 1973. The truck is a 45-ton Terex R-45. *KH*

Figure 4.36 Coal extraction from the steeply dipping seams at Ty-Uchaf is accomplished by a Ruston-Bucyrus 22-RB shovel and some hand labour. *KH*

CHAPTER 4   1970-1985   **121**

Figure 4.37 Terex R-45s dump their load at the Ty-Uchaf site in South Wales. The exhaust-heated body on these trucks connects to the exhaust pipe when in its lowered position, but the exhaust escapes when the body is raised, causing a blackened area on the truck. *KH*

Figure 4.38 In 1971, Taylor Woodrow added this Marion 183-M crawler dragline of nine cubic yards capacity to its extensive fleet of earthmoving equipment at the Royal Arms group of sites near Dowlais, Glamorgan. It was the only one of this model to work in the UK.

Figure 4.39 Most Ruston-Bucyrus electrically-operated 150-RB excavators were equipped as shovels, but a few worked as draglines. An example is this Taylor Woodrow machine, which spent some time at Trecatty in South Wales.

Figure 4.40 A Euclid R-85 truck receives an 85-ton load from a Ruston-Bucyrus 195-B 12-yard shovel. It's one of two installed at Taylor Woodrow's Trecatty site near Merthyr Tydfil, South Wales. Trecatty and its extension ran from 1975 until 1984.

Figure 4.41 Coal seams steeply folded almost to the vertical at Tir-y-Gof site, typify the complicated geology in South Wales. The Tir-y-Gof site at Upper Cwmtwrch was active in the early 1970s. *KH*

CHAPTER FIVE

Walking Draglines

We saw in Chapter 2 how several large walking draglines, some the largest of their type, were imported from America during the 1950s to work on UK opencast sites. Most of these continued to work for at least three decades, including throughout the lean and low-output years of the 1960s, when they were employed on the larger, long-term sites. In 1961, the Opencast Executive (OE) commissioned a Rapier W1800 at Maesgwyn site in South Wales operated by G. Wimpey & Co. Ltd. At 40 cubic yards capacity, the British-built machine briefly claimed the title of the world's largest dragline (see the story of Maesgwyn in Chapter 7). During the ten years following the W1800 installation, only one new walking dragline entered the UK coal industry: that was in 1969 when the 65-yard Bucyrus-Erie 1550-W, named 'Big Geordie', went to work at Radar North site for Derek Crouch (see the Radar North story in Chapter 7).

In 1971, as a taste of what was to come, Wimpey put to work a new Marion 7500 walking dragline in the 20-yard class at its Mabel Plantation site, Workington, Cumbria, following up with an 18-yard Bucyrus-Erie 480-W at the adjacent Outgang site in 1977 where both machines worked. When this site finished in 1983, the two walkers stayed in Wimpey's fleet for a number of years, with the 480-W remaining in the area to complete the Potatopot site (1986–1994) and the 7500 finding work at Chapmans Well, County Durham (1986–1992).

TABLE 5.1 Walking draglines new to UK opencast coal mining since 1970

Make	Model	Size (yard^3)	Date	First site and location	Owner/ Operator
Marion	7500	20	1971	Mabel Plantation, Cumbria	Wimpey
Ruston-Bucyrus	1260-W	32	1977	Coalfield Farm, Leicestershire	BCO/Merriman-Meighan
Ruston-Bucyrus	1260-W	32	1977	Sisters, Northumberland	BCO/Crouch
Ruston-Bucyrus	480-W	18	1977	Outgang, Cumbria	Wimpey
Ruston-Bucyrus	380-W	12	1979	Oughterside, Cumbria	Miller
Ruston Bucyrus	380-W	12	1979	Cadger Hall, Scotland	BCO/Crouch
Ruston-Bucyrus	380-W	12	1980	Tanners Hall, Durham	Lomount
Ruston-Bucyrus	1260-W	30	1981	Headlesscross, Scotland	BCO/Murphy
Rapier	W2000	31	1982	St Aidans, Yorkshire	BCO/Miller
Rapier	W2000	31	1983	E. Chevington, Northumberland	BCO/Crouch
Rapier	W700	13	1983	Godkin, Derbyshire	NSM
Rapier	W700	13	1983	Godkin, Derbyshire	NSM
Ruston-Bucyrus	1260-W	32	1983	Coalfield Fam, Leicestershire.	BCO/Shand
*Marion	7500	20	1989	Nant Helen, Wales	Fairclough-Parkinson
*P&H	757	65	1992	Stobswood	BCO/Crouch

* These machines were added in later years, beyond the scope of this book. A brief description can be found in the Epilogue.

With Britain's 1974 'Plan for Coal' coming into force, the OE and its contractors began looking at the latest designs of walking draglines and placing orders. Between 1976 and 1983, the OE made available to its contractors several walking draglines (see Table 5.1). Six of these were in the 32-cubic yard class, a large size for the UK. Four were the well-proven 1260-W model designed by the Bucyrus-Erie Company in America, but built by Ruston-Bucyrus in England, thus designated Ruston-Bucyrus machines. The other two were also British built by Ransomes & Rapier Ltd. of Ipswich. They were W2000 models, the largest built by the company.

The first 1260-W, named 'Little John' with a 260-foot boom and weighing in at 1,845 tons, was erected at the large Coalfield Farm site at Ibstock, Leicestershire. It worked there from 1977 until 1981, first for Merriman-Meighan Ltd., and later for Shand Mining. The site was a very productive one for the OE; 4.7 million tons of coal were extracted at the favourable overburden-to-coal ratio of 1:4.6, a figure unusually low for the UK.

The second 1260-W ordered by the OE, with identical specification, was erected at the Sisters site near Widdrington, Northumberland. Operated by Derek Crouch Ltd., the site, an extension to the former Radar North site, was completed in 1980. The machine was then transferred to the new West Chevington site operated by A.F. Budge and named Chevington Collier. It stayed there, helping to uncover more than five million tons of coal until 1992. After a major overhaul, the 1260-W spent some three years at Colliers Dean site with the same contractor, now known as R.J. Budge (Mining) Ltd. Finally the machine moved to the nearby Maidens Hall site in 1996 where it worked its last years until being scrapped in 2009.

In 1981, the OE commissioned its third 1260-W, this time in Scotland at the two-million-ton Headlesscross site near Shotts, Lanarkshire, operated by Murphy Brothers Ltd. This machine was specified with a longer 285-foot boom, but otherwise built with the same specification as the first two. In 1989 it moved to the nearby Damside site to work with contractors Currall, Lewis & Martin Ltd. Almost three million tons of coal were recovered from this site up to 2000. After that, the 1260-W sat idle until 2009, when it was dismantled and sold to South Africa.

The fourth 1260-W ordered by the OE was in 1980 for anticipated extension of work at Coalfield Farm site in Leicestershire, and specified identical to the others, except for the longer 285-foot boom. After the usual erection period, it was named 'Big John' and handed over to contractors Shand Mining in 1983, where it joined the first 1260-W still in operation. When the OE received permission to extend Coalfield Farm site to an adjacent area called Coalfield North, both machines were moved there and a further 7.7 million tons of coal were recovered from what was said to be one of Britain's most successful coal mines.

Coalfield North finished in 1992, and with the few expansive, long-term sites in the UK already fully equipped with suitable draglines, the OE decided to put the two 1260-Ws up for sale. After a long and intensive sales effort, no serious buyer could be found and they were scrapped in 1998. This wasteful action appears inconceivable for these well-maintained, low-hour machines in perfect working order, especially the one built in 1983 with a mere nine years of work tallied in its log book. Even more remarkable is the fact there are draglines working today in Pennsylvania, USA that were built in the mid-1940s! A walking dragline is normally expected to enjoy a working life of at least 30 years.

The first of the two Rapier W2000s went to work at Miller Mining's large St Aidans Extension site near Leeds, Yorkshire in 1982. It spent its entire life there, including a long idle period when the site was flooded, and was finally scrapped in 2004. The other W2000 was assigned to Crouch Mining for its East Chevington site in Northumberland, and started digging in 1983. Like its sister machine it never moved from its original location, helping to uncover 3.8 million tons of coal until 1993. After a brief idle period, this machine was scrapped in 1995. The two Rapier W2000s weighed approximately 2,000 tons and were of identical specification, with 31-yard buckets suspended from 314-foot booms.

Again, it seems strange that these two relatively new Rapier machines in good condition would be scrapped, each with little more than ten years of work behind them. Even stranger is the fact that Rapier continued to build this same model throughout the 1980s for export to India, and even shipped some after the East Chevington machine was scrapped. Possibly the financial institution behind these Indian investments insisted on only new equipment?

Smaller walking draglines also found their way into UK opencast sites beginning in 1979. These included three Ruston-Bucyrus model 380-Ws, the first of a new breed of dragline designed on modular principles, bolted together so they could be dismantled easily and moved in 'roadable' components. The first, diesel-powered by a Caterpillar 12-cylinder engine and swinging a 12-yard bucket on a 170-foot boom, was purchased by the OE and allocated to Crouch Mining at Cadger Hall, Scotland. Miller Mining took the second 380-W for its Oughterside site in Cumbria. Unlike the first, this machine was all-electric, taking power from a trailing cable, but otherwise of similar specification and bucket size as the first. On completion of Oughterside, Miller's 380-W saw work at Acton Extension in Yorkshire (1986–1990) and Kirk in Derbyshire (1990–1998).

The third 380-W, another electric machine, was delivered in 1980 to Lomount Construction Ltd. for its Tanners Hall site in County Durham. On completion of this site in 1986, the 380-W was transferred to Northern Strip Mining/Coal Contractors (NSM) for four years' work at Ponesk site in Scotland, then another five years at Rainge near Clay Cross in Derbyshire. The advantage of an on-board diesel engine might have been appreciated here, because this electric machine was operated from an inconvenient portable diesel-electric generator in an effort to avoid installation of an expensive power distribution system on site. After Rainge, the 380-W stayed with NSM and moved back up north for a further four years work at Low Gordon site in County Durham.

The smaller, modular-designed walking dragline concept was picked up by Ransomes & Rapier Ltd. with the introduction of the W700. After exporting two to C&K Coal Company in America in 1981, and installing one with London Brick Company, Rapier sold two to NSM for its Godkin site in Derbyshire. Commencing in 1983, and working alongside the O&K RH-300, the world's largest hydraulic excavator, the 13-yard machines did not last the duration of this site, and by 1988 they were idled and advertised for sale.

Figure 5.1 Wimpey's Marion 7500 walking dragline at Mabel Plantation site in Cumberland was one the first of this type to be imported from America and one of only two to operate in the UK opencast coal industry. The site called for the removal of 32 million cubic yards of overburden and delivery of 1.8 million tons of coal from 1970 to 1979. *KH*

Figure 5.2 The Marion 7500 at Mabel Plantation began operation in late 1971. The all-electric 765-ton machine worked with a 17-cubic yard bucket on a 200-foot boom, and was equipped with a single DC main motor of 1,000-horsepower to operate its hoist and drag drums through clutches and brakes. *KH*

Figure 5.3 Definitely a dragline job! An overall view of Mabel Plantation site showing two Lima 2400 draglines in the foreground. In the distance, the Marion 7500 dumps another 17-yard bucket load on the spoil piles. *KH*

CHAPTER 5 WALKING DRAGLINES **129**

Figure 5.4 Depending on the mine plan and equipment available, loading haul trucks with a dragline is sometimes necessary when the material must be hauled some distance. This might occur when box cut material is required in distant locations or fill material is needed to construct haul roads or dams. Here, a Terex R-50 takes on a full load from a Lima 2400 dragline at Mabel Plantation. *KH*

Figure 5.5 Hand labour is much in evidence here while the 22-RB coaling shovel gleans thin seam coal from a steeply-dipping face. Labourers collect every last piece of coal and ensure stray rocks are removed. *KH*

Figure 5.6 The was the first 1260-W ordered by the Opencast Executive. Named 'Little John', it worked at Coalfield Farm site from 1976 until 1981, initially for Merriman-Meighan Ltd. and later for Shand Mining. The 1,845-ton machine then moved to the adjacent Coalfield North site where it was joined by a second 1260-W, 'Big John', and worked until 1992. These sites yielded a combined 12 million tons of coal.

Figure 5.7 This is how the 1260-W was operated at Coalfield Farm site. Called a 'chop cut', the dragline first digs above its own level and removes the upper section of overburden and swings this approximately 180 degrees to dump into the previously mined out cut. It then moves forward to remove the remaining overburden to expose the coal, swinging this approximately 90 degrees into the previous cut. Ruston-Bucyrus 22-RB and 38-RB shovels load coal in the foreground. *KH*

CHAPTER 5   WALKING DRAGLINES   131

Figure 5.8 Cable and hydraulic excavators excavate and load coal in this scene at Coalfield Farm. They are aided by a Caterpillar wheel loader and a D9 crawler tractor with Kelly ripper. *KH*

Figure 5.9 The third 1260-W purchased by the OE began life in 1981 at Headlesscross operated by Murphy Brothers Ltd. This photo (2003) shows the 1260-W in its final resting place before being sold, after completing its work at the nearby Damside site, Lanarkshire, Scotland for contractors Currall, Lewis & Martin Ltd. *KH*

Figure 5.10 This Rapier W2000 dragline, the first of two purchased by the OE in 1982, is shown working at Miller Mining's large St Aidans Extension site near Leeds, Yorkshire. With a 314-foot boom and 31-yard bucket, the walker spent its entire life there and was finally scrapped in 2004. *KH*

CHAPTER 5   WALKING DRAGLINES   133

Figure 5.11 A smaller walking dragline was the Ruston-Bucyrus 380-W, built on modular principles and relatively easy to move from site to site. It found use in certain opencast coal applications and carried a 12-yard bucket on a 170-foot boom. The first was this one at Cadger Hall, Scotland, operated by Crouch Mining. It was powered by a Caterpillar 12-cylinder diesel engine.

Figure 5.12 This 380-W, shown at Coal Contractors' Rainge site in Derbyshire, is a straight electric machine, fed by trailing cable from a portable diesel generator. In a multiple coal seam operation, a dragline was often useful to remove the bottom layer of overburden (interburden) that could be dumped directly into the worked-out pit. *KH*

Figure 5.13 Another modular-designed walking dragline of smaller proportions was the Rapier W700. Northern Strip Mining Ltd. purchased two of these for its Godkin site in Derbyshire. Commencing in 1983, the machines worked with 13-yard buckets on 165-foot booms. *KH*

Figure 5.14 The second Ruston-Bucyrus 1260-W ordered by the OE started work at the Sisters site near Widdrington, Northumberland, in 1976. After several moves to sites in the same area, the big walker named Chevington Collier started work at Maidens Hall site in 1996, where it worked its last years until it was scrapped in 2009. *KH*

CHAPTER SIX

Hydraulic Excavators

The 1970s saw the introduction and wide acceptance of a new form of excavating machine into opencast coal mining, the hydraulic excavator. The trend started in 1970 when Northern Strip Mining Ltd. (NSM) installed two medium-sized machines of four cubic yards capacity at its Gresley site at Swadlincote. That same year, French manufacturer Poclain, an early pioneer of the hydraulic excavator, displayed the world's largest at the Intermat equipment show in Paris. Grabbing trade headlines around the world, the Poclain EC1000 signalled the advent of a new era of hydraulic machines for bulk excavation, one that would revolutionize the British opencast coal scene.

After almost two decades of development, gradual increases in size, sophistication and reliability, hydraulic excavators were now finally able to challenge the domain of large cable excavators in surface mining. The EC1000 boasted an operating weight of 160 tons, and was powered by no fewer than three GM 8V71 diesel engines developing 840 horsepower. The independent crawler drive and tractor-type crawlers and sprockets endowed this monster machine with remarkable agility.

The advent of hydraulic excavators in the 1940s laid the foundation for a complete revolution of the excavator industry. The 1950s were the pioneering years when a handful of manufacturers ventured into building this new type of machine, but unreliable hydraulic systems unable to withstand the rigours of excavation slowed development. Germany's Demag produced the world's first 360-degree slew, crawler-mounted hydraulic excavator in 1954, the B504 of ½-cubic yard capacity, but few were sold. During the 1960s, hydraulic excavators increased in size, and more manufacturers entered the field. They developed more reliable, higher pressure hydraulic systems initially successful on the early European machines. Their persistence paid off, and by the end of the 1960s, hydraulic excavators had graduated into a major force in the excavating industry.

The arrival of Poclain's EC-1000, moving up to the record-beating ten-yard class, was regarded with apprehension by traditional cable excavator manufacturers. Here was a machine encroaching well into territory previously believed to be the sole domain of cable machines. Concerns were justified, as towards the end of the next decade excavators of much greater power and capacity were marketed by a few dominant manufacturers. By 1980, the O&K RH-300 was already at work, an excavator with a bucket capacity of 34 cubic yards, three times the size of the Poclain EC1000.

Contractors working on British opencast coal gradually embraced this new type of excavator. NSM led the way in 1970 when it imported the two O&K RH-25 excavators from Germany, and put them to work on the company's major Gresley site. These four-yard clamshell-type shovels were teamed with a fleet of Terex R35S dump trucks. The excavators were so successful that within two years, NSM purchased the newest and largest O&K excavator available, the 8½-yard RH-60 with a service weight of 120 tons. This was placed at the firm's new Acrefair site near Wrexham, North Wales in 1972, where removal of 52 million cubic yards of overburden would yield two million tons of coal and more than five million tons of fire clay. Before the site finished, no fewer than four O&K RH-60s were working there and NSM had placed a fifth at its Gresley site.

In 1972, Miller purchased the first Poclain EC1000 excavator in England for its Pit House site near Sheffield. This world's largest hydraulic excavator worked as a 9½-yard shovel and was teamed with a fleet of 45-ton Terex R-45 trucks. The site was also home to a Lima 2400 shovel and an eight-yard Manitowoc 4600 dragline. After only nine months' work with the first EC1000, Miller committed to a second machine for its St Aidans site near Leeds. In 1974, Miller's shovel choice for a newly won contract at Albert Colliery near Wigan, Lancashire was another Poclain. This time it was the uprated 1000CK, now with its weight increased to 180 tons, bucket capacity to 11½ cubic yards and with twin Deutz engines totalling 900 horsepower. A year later, Miller added an O&K RH-60 shovel to its extensive fleet at St Aidans. This site is described in more detail in Chapter 7.

By the time O&K upgraded its RH-60 to the 150-ton RH-75 in 1976, NSM's proven success with large hydraulic excavators had attracted much attention from others in the industry. Several contractors, including Miller Mining, were eager to experience the new ten-yard RH-75, and orders steadily flowed in over the next few years with NSM and Miller taking five each. When O&K introduced the upgraded RH-75C in 1983, further orders were received from Taylor Woodrow, Budge, Miller and Bayford Mining.

But O&K and Poclain did not have the UK hydraulic excavator business all their own way, as the 1970s saw several other manufacturers making their debut. In 1978, Germany's Demag sold two 120-ton H111 shovels with eight-yard buckets, Demag's largest at that time, to Fairclough-Parkinson for work at Anglers site in Yorkshire and Blindwells site in Scotland. This company also employed an 80-ton Hitachi UH-30 with six-yard bucket at Anglers. Shand Mining purchased a 134-ton Koehring 1266D with a 6½-yard backhoe for its Park Slip site in South Wales, to join a Poclain 1000CK. A second 1000CK went to the company's Rhigos site. In 1980, contractor D.P. McErlain (later Coal Contractors) brought the first of Demag's newest and largest excavator into the UK. The H-241, weighing in at 262 tons and wielding an 18-yard bucket, worked at the Ryefield site at Denby, Derbyshire. And through hire company Short Brothers, larger 13-yard Hitachi EX1800 and 23-yard EX3500 excavators were placed on sites throughout the UK.

Compared with cable types, hydraulic excavators are cheaper to buy but more expensive to operate. However their advantages made them particularly suited to UK conditions:

- Their breakout force at the bucket teeth is greater than a cable shovel, and the prising force from the bucket's 'wrist' action allows harder material to be excavated without drilling and blasting. This is a big advantage in the UK where sites are often close to residences.
- Positive digging action enables the bucket to penetrate the face at any height, allowing selective digging and easier handling of blocky material.
- Independent hydraulic power to each crawler track provides superior mobility and travel speeds are higher.
- Although a few machines are supplied with electric power for special applications, by far the greater majority are diesel-powered. This eliminates the need for a trailing cable and costly site electrification, a major advantage in the UK where sites are often of short duration.
- Machines are designed in modules for quick and easy disassembly when moving from site to site.
- Down pressure from the hydraulic front end allows a hydraulic machine to lift its crawler tracks off the ground. This is useful for loading a machine onto a trailer or extricating itself from soft material.

With all these advantages in their favour, and the fact that by 1980 they had progressed to sizes at least as great as the largest cable excavators employed in the UK, it became obvious that hydraulic excavators would play a most significant role in the future of British opencast coal.

And the influx of hydraulic excavators into opencast coal was by no means restricted to large overburden removal machines. Dozens of smaller hydraulic excavators for coal cleaning and coal loading rapidly replaced traditional cable machines such as the Ruston-Bucyrus 22-RB. O&K's smaller hydraulic machines such as the one-yard RH-9 were popular, and Caterpillar, who first entered the hydraulic excavator field in 1972, shipped large numbers of its biggest model at that time, the 245 model with five-yard shovel or 3½-yard backhoe. Hydraulic excavator manufacturers listed in a 1981 OE report were Akerman, Caterpillar, Demag, Ford, Hitachi, Hymac, International, Deere, JCB, Komatsu, Liebherr, Massey-Ferguson, O&K, P&H, Poclain, Priestman and Ruston-Bucyrus.

The 1970s ended on a high note when Northern Strip Mining Ltd. (NSM) broke another record for size when it purchased the world's largest hydraulic excavator in 1980, the RH-300 from O&K. With an operating weight of 535 tons, a 34-yard bucket and power from two 1,210-horsepower Cummins KTA2300C diesel engines, it easily surpassed any hydraulic excavator built up to that time. Surprisingly, the massive machine, built on similar lines to O&K's smaller models, was able to move easily from site to site in modular components. It first spent a short time at NSM's Donnington Extension site near Burton-on-Trent before being moved to the company's large Godkin site near Heanor, Derbyshire, which ran from 1981 to 1990. This RH-300 was the only one to operate in the UK.

In the early 1980s, Liebherr machines from Germany also appeared, including an eight-yard 982 and 13-yard 991 in action for Merriman-Meighan Ltd. at Natsfield site near Cannock. These were followed by 195-ton R994 excavators with 15-yard buckets purchased by Clay Colliery Co. Ltd., Kier Mining and Wimpey. In 1981–1982, Simms Sons & Cooke imported five 175-ton P&H 1200 excavators built in Germany. These 15-yard machines were employed at Simms' Millers Lane and Amberswood sites in Lancashire and Pica in Cumberland. The ultimate in size for a hydraulic excavator in the UK was reached in 1986 when Coal Contractors put to work the huge Demag H485 at its Roughcastle site near Bonnybridge, Scotland. Slightly heavier than the O&K RH300 at 620 tons, it carried a 34-yard bucket and was powered by a single MTU engine delivering 2,106 horsepower.

The most popular of all the large hydraulic excavators in the UK was the O&K RH-120C. The first was purchased by Budge in 1983 and went to the Kinswood site near Cannock. The 15-yard shovel or backhoe proved to be a perfect match for the 100-ton trucks (Caterpillar or Terex) that were now becoming very popular. These teams of excavator and trucks were also well-suited to the three- to six-year average duration of most UK sites. Before long, almost every British opencast coal contractor owned at least one RH-120C and many boasted multiple machines in their fleets. The winner was Crouch, proud owner of 22 in its fleet, followed by Miller with 20.

As the number of hydraulic excavators increased rapidly, the industry experienced a corresponding decrease in small and mid-sized draglines and cable shovels. During the 1980s the total number of Lima and Manitowoc diesel machines fell by about 80 per cent, the number of Ruston-Bucyrus 150-RB electric shovels decreased by 50 per cent and 110-RBs and 71-RBs became non-existent. Over the same period, smaller rope excavators used for coal cleaning and loading were almost eliminated.

138 CHAPTER 6 BRITISH OPENCAST COAL

Figure 6.1 Northern Strip Mining Ltd. (NSM) led the British opencast coal industry into use of hydraulic excavators for bulk excavation and overburden removal. In 1970, it placed two of these O&K RH-25 shovels on its Gresley site near Swadlincote where the four-yard machines were so successful that much larger hydraulic excavators soon appeared. *KH*

Figure 6.2 A view of NSM's Gresley site at Swadlincote. On this multiple-seam operation, a variety of machines remove overburden and coal, including a 3½-yard Ruston-Bucyrus 71-RB shovel, an O&K RH-25 shovel, Terex R-35S dump trucks and a coal transport lorry. *KH*

Figure 6.3 An electric-powered Ruston-Bucyrus 110-RB shovel loads a Terex R-45 truck for NSM at Gresley. In the background, Hymac and O&K excavators load coal. *KH*

Figure 6.4 Scrapers always played an important part on NSM's sites. Here a mixed fleet of Terex and Caterpillar remove sections of overburden at Gresley site. *KH*

Figure 6.5 Hand work to this extent typifies opencast coal in the UK. Mining thin seams that would be discarded in other countries means that every last inch of the high-quality coal must be gleaned. Working with high-production excavators, even when fitted with special toothless buckets, still requires the delicacy of a hand shovel to clean the surface. *KH*

Figure 6.6 In 1972, Miller Mining was first to import a Poclain EC1000 into the UK, shown here at Pit House opencast site near Sheffield. The 160-ton machine with 9½-cubic yard bucket was the world's largest hydraulic excavator and was powered by three engines putting out 840 horsepower. *KH*

Figure 6.7 During the 1970s, hydraulic excavators encroached into all areas of excavation formerly carried out by cable machines. Here at Pit House, a Hymac 880C with special flat-bottomed face shovel loads the coal. *KH*

Figure 6.8 Making trade news headlines in 1972, NSM imported the first RH-60 shovel from Germany and put it to work at the firm's Acrefair site near Wrexham, North Wales. By 1975, four of these 120-ton machines had seen service at Acrefair, and another – shown here loading a 50-ton Caterpillar 773 truck – was placed at the company's Gresley site. *KH*

142 CHAPTER 6 BRITISH OPENCAST COAL

Figure 6.9 In 1976, O&K upgraded its RH-60 to the 150-ton RH-75 and contractors were quick to order the new ten-yard machine. Over the next few years many contractors, such as Taylor Woodrow who operated three, added them to their fleets. This one worked at its Butterwell site. *KH*

Figure 6.10 When introduced in 1978, the Demag H-241 was claimed to be the world's largest hydraulic excavator at 262 tons operating weight. This one, with 18-yard backhoe bucket, loads Euclid R-85 haul trucks at Miller's St Aidans site near Leeds in 1986. *KH*

CHAPTER 6 HYDRAULIC EXCAVATORS   143

Figure 6.11 The O&K RH-9 was one of the most popular coal-loading excavators. This one worked at Budge's Arkwright site in Derbyshire. Notice the special dipper designed for accurate loading of thin seams, but still needing hand work to ensure a clean product! *KH*

Figure 6.12 Another popular mid-size machine, widely used in the UK for overburden removal, was the Caterpillar 245 excavator, either in shovel or backhoe form. This shovel at NSM's Gresley site, with a five-cubic yard clam-type bucket, is shown loading a 50-ton Terex R-50 dump truck. *KH*

144 CHAPTER 6 BRITISH OPENCAST COAL

Figure 6.13 Terex trucks, named Euclid prior to 1968, were very popular throughout the UK because they were made in Scotland. This R-35S is one of NSM's fleet at its Gresley site. *KH*

Figure 6.14 NSM broke another record for size when it purchased the world's largest hydraulic excavator in 1980, the O&K RH-300. The only machine if its type in the UK is shown loading a 170-ton Wabco dump truck, also the largest in the UK, at Godkin site near Heanor, Derbyshire in 1981. *KH*

Figure 6.15 Another view of the giant RH-300. The 535-ton excavator could load the 170-ton truck in four passes with its 34-yard bucket.

Figure 6.16 The four Wabco 170C haul trucks at Godkin site in Derbyshire were the only ones of this type to work in the UK. The diesel-electric units were rated at 170 tons capacity and powered by either Detroit or Cummins engines providing 1,600 horsepower. *KH*

146 CHAPTER 6 BRITISH OPENCAST COAL

Figure 6.17 O&K not only made a major impact in the British opencast industry with its excavators, it also sold a few of its large 77,000-pound graders, heavier than most at the time. This one, photographed in 1981 at Godkin site, maintained smooth haul roads for NSM's Wabco 170-ton haulers. *KH*

Figure 6.18 Liebherr excavators also enjoyed a certain amount of success in the UK. This 195-ton R994 with 15-yard bucket worked for Clay Colliery Co. Ltd. at its Ketley site near Telford.

CHAPTER 6   HYDRAULIC EXCAVATORS   147

Figure 6.19 Two P&H 1200 shovels remove overburden at Millers Land site near Wigan, operated by Simms, Sons & Cooke Ltd. This company imported five such machines from Germany for their various sites. *KH*

Figure 6.20 The largest hydraulic excavator ever seen in the UK is this 620-ton Demag H-485 put to work by Coal Contractors at Roughcastle site near Bonnybridge, Scotland, in 1986. It carried a 34-yard bucket and was powered by a single MTU engine developing 2,106 horsepower. *Euan Ramage*

Figure 6.21 Most popular of all the large hydraulic excavators in the UK was the O&K RH-120C. This one, belonging to UK Coal (Formerly R.J. Budge) is busy at the Orgreave site, Sheffield. Equipped as a 15-yard shovel or backhoe, the RH-120C was a perfect match for the 85 to 100-ton Caterpillar and Terex dump trucks. *KH*

Figure 6.22 In 1989, Budge purchased the first RH200, which was shipped to the West Chevington site, Northumberland, after first appearing at that year's BAUMA show in Germany. Seen here is Budge's second RH200 at Orgreave site near Sheffield. At 440 tons operating weight and with a 26-yard bucket, it's a good match for the Caterpillar 789 195-ton hauler. *KH*

Figure 6.23 Activities at the Alexandra site in 1987 near Wigan operated by Raymin Northern Ltd. shows hydraulic excavators doing most of the work. In the foreground a JCB805B cleans the coal. Mid-distance a Demag H-85 removes final overburden and loads a Euclid R-50 truck. In the background a Demag H-185 takes the upper overburden. *KH*

Figure 6.24 'Bell' pits exposed in ancient underground coal workings. Hydraulic excavators are the most efficient tool to excavate these cleanly… perhaps with a little hand work as well! Old underground workings were not appreciated by opencast crews. Once exposed, they were expensive to clean and increased the overburden-to-coal ratio. *KH*

Figure 6.25 All manufacturers of large hydraulic excavators, most notably German and Japanese, enjoyed a prolific market for their products in the UK. This Demag H-185, belonging to UK Coal, works at Park Brook site near Sheffield. It loads a new Terex TR-100 truck. *KH*

CHAPTER SEVEN

Significant sites

This chapter covers in detail ten major opencast coal sites in the UK. Of course not all could be included in a book of this size, but those selected feature a good cross-section of interesting sites in England, Scotland and Wales. Their size, longevity or use of special equipment qualify them as significant. They are described here in alphabetical order.

Abercrave – South Wales

In 1964, Derek Crouch (Contractors) Ltd. was awarded the large Abercrave opencast coal contract located close to Abercrave village, some 20 miles north-east of Swansea, South Wales. Crouch was already well-established, with opencast activities in the north-east of England where it held major contracts for several sites including the prestigious Radar North group. So it was interesting to see Crouch expand into the South Wales coalfield largely dominated by the likes of Wimpey and Parkinson. The Abercrave contract called for the delivery of 2.3 million tons of coal over a nine-year period. Although this doesn't seem a large amount of production for that time span, the complicated geology and overburden-to-coal ratios between 25 and 30 to one, made this a significant undertaking.

Work began in earnest to move and re-erect the largest excavator planned for Abercrave. This was one of the four Bucyrus-Erie 1150-B walking draglines employed in the UK on opencast coal-mining. Two of these worked from 1954 to 1963 at the Tirpentwys site near Pontypool for Wilson Lovatt, and one of these was moved to Abercrave. The other was moved to the Poplars site near Cannock, Staffordshire.

When introduced, the 1150-B was one of the largest draglines in the world. The Abercrave machine swung a 27-yard bucket on a 180-foot boom, providing a digging reach of 174 feet with the boom set at 30 degrees. The circular base measured 44 feet in diameter and overall machine width was 63 feet. Tail swing radius was 50 feet six inches and working weight was quoted as 1,230 tons. DC electrical equipment included two hoist motors, two drag motors, three swing motors and two walk motors. These were driven by the Ward-Leonard control system from two AC motor generator sets totalling 1,750 horsepower.

Crouch fielded a varied line-up of equipment for the Abercrave site. In the early years, a Rapier W150 walking dragline with six-yard bucket uncovered outcrop and shallow coal reserves along the perimeter of the site. The big 1150-B stayed at the lower levels of the site where deeper overburden was encountered, while the upper levels were handled by two Ruston-Bucyrus 150-RB electric shovels with dippers rated at six cubic yards capacity, purchased new for this contract.

To serve the 150-RB shovels, Crouch invested in a fleet of 35-ton Caterpillar 769 dump trucks, one of the first such fleets to operate in the UK. Announced in 1962, the 769 was Caterpillar's first off-highway truck and was powered by a Caterpillar D343 diesel engine developing 400 flywheel horsepower. Gross vehicle weight with 35 tons payload tipped the scales at 125,920 pounds. As haulage distances increased, the 769s were supplemented by a fleet of Aveling-Barford SN35 dump trucks of similar capacity.

Other equipment included a brand new 824 wheel dozer to keep shovel loading areas tidy and free from rocks, and its relatively high speed allowed it to move quickly to other assignments such as tip maintenance. Since the 300-horsepower dozer was only released by Caterpillar in America the year before Abercrave started, it likely was one of the first in the UK. Other Caterpillar machines on site included D8 bulldozers, and a model 955 crawler loader and 944 wheel loader to assist with coal cleaning and loading. Other machines in the coal-loading department included Ruston-Bucyrus 22-RB and 30-RB shovels that loaded directly into lorries, which hauled to the nearby NCB disposal point.

After the Abercrave site was completed in 1973, the 1150-B dragline was moved to the nearby Onllyn site operated by Wimpey from 1974 until 1982. After that it was dismantled and sold to Beltrami Enterprises, Inc., Pennsylvania, USA.

Figure 7.1 The Bucyrus-Erie 1150-B walking dragline at Abercrave, one of four imported from America by the Department of Opencast Coal Production. *KH*

Figure 7.2 The 27-yard bucket on the 1150-B undergoes a spot of welding. *KH*

154 CHAPTER 7 BRITISH OPENCAST COAL

Figure 7.3 The 1150-B dragline excavated the lower reaches of the Abercrave site while 150-RB electric shovels removed the upper layers of overburden. An Aveling Barford SN35 negotiates the ramp toward the tip area. *KH*

Figure 7.4 One of two Ruston-Bucyrus 150-RB electric shovels, purchased new for the Abercrave contract, loads a Caterpillar 769 dump truck. *KH*

CHAPTER 7 SIGNIFICANT SITES 155

Figure 7.5 A busy scene at Abercrave shows the main cut and coal-loading activities in 1967. *KH*

Figure 7.6 A 150-RB electric shovel loads the top bench of overburden at Abercrave. The fleet of Caterpillar 769 dump trucks was one of the first in the UK. *KH*

Figure 7.7 Another 'first' for Crouch at Abercrave was this Caterpillar 300-horsepower 824 wheel dozer. *KH*

Figure 7.8 One of Crouch's 22-RB shovels loads anthracite into a road-going lorry. *KH*

Acorn Bank – Northumberland

In 1955, the Opencast Executive (OE) of the National Coal Board (NCB) awarded a contract to Costain Mining Ltd. for the production of five million tons of coal over a seven-year period from the Acorn Bank site at Bedlington, Northumberland. Later extensions yielded a further two million tons of coal, and the site didn't actually finish until 1966. Worldwide interest was attracted with the opening of the box (first) cut, the largest man-made hole in Europe. Visitors, including mining engineers and technicians, reportedly travelled from more than 20 countries including Russia and the United States to see excavation taking place on so vast a scale.

Equipment engaged on overburden removal included Joy-Sullivan rotary blast hole drills, two Bucyrus-Erie 1150-B walking draglines with 25 cubic yard buckets owned by the OE, three five-yard Ruston-Bucyrus 120-RB electric shovels, two 3½-yard Lima 1201 shovels and a fleet of more than 60 Euclid rear dump trucks. One of the 1150-Bs, purchased from Parkinson, was walked from the nearby Ewart Hill site. The OE purchased the other one from Illinois, USA. These 1,200-ton 1150-Bs were two of four employed by the OE in the early 1950s. The other two worked at Tirpentwys in South Wales and were the largest excavators to work in the UK at that time.

The box cut, some 230 feet in depth and involving excavation of ten million cubic yards of rock and earth, was completed in less than a year. While this in itself was quite an achievement, even more remarkable is the fact that six million cubic yards of this material were hauled by Euclid trucks to backfill the last cut at the Ewart Hill site. The other four million yards were hauled to storage tips for replacement during restoration. While all this box-cut material was being hauled away by the 60-strong Euclid fleet, the big 1150-B draglines were called upon to assist the three 120-RB shovels with loading the trucks. Thus the 22-ton Euclids were loaded with 25-yard dragline buckets! Certainly not a normal situation, but nevertheless achieved successfully with the aid of a home-made skid-mounted hopper/chute structure to reduce spillage, not to mention skilful operators!

By June 1957, the site settled into its normal operating sequence, and average production of 16,000 tons per week was attained. Peak weekly production reached 20,800, claimed to be a world record from a deep opencast site. Coal excavation was handled by a fleet of Ruston-Bucyrus, NCK and Lima shovels ranging in size from ¾-yard to two-yard capacity.

Another 'first' claimed at Acorn Bank was the use of four high-speed derrick cranes to hoist coal from the bottom of the cut. The ten-ton skips were dumped into an eight-strong fleet of 40-ton Euclid bottom-dumping tractor-trailer units, ordered specially for this contract and claimed to be the largest of their type in Europe. The rail-mounted electric derricks, manufactured by Butters Brothers, were fitted with 152-foot long booms. They could hoist a load of 13 tons at 140 foot radius at a speed of 120 feet per minute. When installing the Butters derricks, particular attention was paid to operator visibility. The cab was located on the mast, some 36 feet above rail level to provide an uninterrupted view of the skip. Skilful operators positioned the skips adjacent to coal shovels at the bottom of the cut, almost 200 feet below their eye-level. When thick coal seams were encountered, the derricks operated with five-ton capacity grabs that excavated coal directly from the seam and loaded coal haulers on the bench. A track-laying crew was employed to look after the derrick rail system. Operation of these derricks eliminated the high cost of constructing ramps for hauling the coal up them to the surface.

Topsoil and subsoil was salvaged using a fleet of Caterpillar D8 and Vickers VR-180 crawler tractors with pull-type scrapers. Ahead of the draglines, the three 120-RB electric shovels worked in 45-foot deep benches to remove the upper overburden, which was loaded into trucks and hauled onto the spoil piles. This reduced the need to level the dragline spoil piles as they were covered with the upper overburden material. The shovel and truck operation formed a wide, level operating bench from which the draglines, overburden trucks and rail-mounted coal derricks operated. The two draglines, one at each end of the cut, removed the remaining overburden and dumped into the previously worked-out cut.

A horizontal coal auger was also employed at Acorn Bank, believed to be the first on British opencast sites. It was used to extract coal from the high walls at the northern and southern limits of the site. Built by Joy-Sullivan, the electrically powered machine was fitted with a 30-inch diameter cutting head and a series of four-foot long auger flights. Holes were drilled to a depth of 60 feet at three-foot centres and about 12 tons of coal were extracted from each hole.

Acorn Bank occupied about 500 acres of agricultural land, which was gradually reclaimed as the site progressed. About 12 inches of topsoil and three feet of subsoil were separated and replaced on the reclaimed land, which returned to agriculture. The area also formed what is now the Bedlington Golf Course next to A1068 Hartford Road.

Figure 7.9 One of the Bucyrus-Erie 1150-B walking draglines with a 25-yard bucket excavates the main cut below the working bench prepared by the shovels. Two Joy-Sullivan drills and two Lima shovels can be seen preparing the bench. The proximity of the local houses in the background would allow a perfect view for the residents!

Figure 7.10 A view of the main cut showing coal loading operations and an 1150-B dragline in the background.

160 CHAPTER 7 BRITISH OPENCAST COAL

Figure 7.11 This view shows the original box cut, up to 230 feet deep, involving the removal of ten million cubic yards in less than a year. The coal hoisting by derricks had not yet been set up when this picture was taken.

CHAPTER 7   SIGNIFICANT SITES   161

Figure 7.12 The 1150-B with 25-yard bucket carefully loads a 22-ton Euclid haul truck. Occasionally the trucks were directly loaded by the bucket without the aid of the mobile hopper.

Figure 7.13 The home-made mobile hopper/chute structure makes it a little easier for the dragline operator to spot the 25-yard bucket and drop its load into the 22-ton truck. No time to rest for the truck driver! One bucket load and he was on his way to the tip.

162 **CHAPTER 7** BRITISH OPENCAST COAL

Figure 7.14 This view shows all four of the Butters derricks on the high wall. They could lift 13 tons at 140-foot radius. One of the 1150-B draglines is seen in the background.

Figure 7.15 The ballasted bogie and 66-inch gauge rail layout is shown on one of the Butters derricks as it hoists coal near one of the draglines. The high position of the derrick cab is necessary for the operator to see to the bottom of the cut, almost 200 feet below.

CHAPTER 7 SIGNIFICANT SITES 163

Calow Herne – Derbyshire

Major road materials and quarrying company, Tarslag Ltd. was very active in opencast coal in the 1950s. Its largest site was Calow Herne, just east of Chesterfield in Derbyshire. Named after the adjacent Herne House, the site started in 1951 and continued well into the early 1960s. Here, Tarslag employed many small to medium-sized excavators, particularly American-built Lima 802 and 1201 shovels and draglines, and Caterpillar D8 crawler tractors. It also ran a large fleet of Euclid 15-ton rear dump haul trucks, no doubt mostly built in Scotland by Euclid (Great Britain) Ltd. at its Newhouse, Lanarkshire factory that opened in 1950.

Only one year after entering the site, Tarslag had stripped topsoil and subsoil from a vast area and several pits had been opened up. Shallow coal seams were stripped entirely by Lima 1201 draglines, while deeper areas were excavated by shovels and the material hauled to external tips. As work progressed over the next few years, additional areas were added and existing pits deepened to mine multiple coal seams. The use of many relatively small machines was actually a benefit in this area of complicated geology where steeply dipping, multiple coal seams could be extracted efficiently.

Throughout the project, Tarslag kept current with its site restoration. Shallow coal areas mined during the initial phases of the site were soon reclaimed and returned to farmland. The first deep pit, after coal extraction, became a depository for overburden from adjacent pits and was progressively filled to original ground level. Subsequent pits were similarly filled, sometimes involving long hauls. Topsoil and subsoil, initially salvaged and stored in mounds around the site by scrapers, was replaced in its natural layers ready for farm cultivation. In 1959, Tarmac Ltd. took over Tarslag Ltd. and earthmoving and construction equipment were amalgamated with Tarmac's existing earthmoving fleet, which continued to work on opencast coal contracts for just a few more years.

The upgraded A617, the main link between Chesterfield and Junction 29 on the M1 motorway, runs through the former site at Corbridge. Green agricultural land, mature trees and some recent housing developments are all that can be seen today, showing no evidence at all of the intense earthmoving activity that extracted hundreds of thousands of tons of much-needed coal at Calow Herne.

Figure 7.16 These two Lima excavators – a shovel and a dragline – are opening up Calow Herne, which would become Tarslag's largest opencast coal site. Initial excavation required long hauls so additional dump trucks were temporarily hired, such as this Euclid R-15 of Euclid Wagon Hirers, Northampton.

Figure 7.17 Tarslag employed a large fleet of Caterpillar D8 tractors for removing topsoil and subsoil for storage during mining. Extensive areas were pre-stripped prior to excavation by shovels and draglines.

Figure 7.18 Caterpillar D8s were mostly paired with Onions scrapers made by Onions & Sons (Levellers) Ltd., later sold by Vickers Armstrong to go with its crawler tractors.

CHAPTER 7 SIGNIFICANT SITES   165

Figure 7.19 At shift change, part of Tarslag's earthmoving fleet lines up ready for refuelling. Euclid R-15 haul trucks are seen with several Caterpillar D8 tractors and pull-type scrapers.

Figure 7.20 Five excavators are seen here, three Lima and two Ruston-Bucyrus, opening up the first cut at Calow. The material is being carried on a long haul to a temporary tip for storage until the final cut in this part of the site is ready for backfilling.

Figure 7.21 A 2½ yard Ruston-Bucyrus 54-RB diesel shovel loads 15-ton capacity Euclid R-15 trucks at Calow Herne. A stockpile of surface soil can be seen on the right.

Figure 7.22 Tarslag fielded an extensive fleet of dump trucks at Calow. Here is one of many Foden 6 × 4 types with a capacity of nine cubic yards.

CHAPTER 7 SIGNIFICANT SITES   167

Figure 7.23 With so many shovels and trucks on the haul road during the initial stages of Calow, the dump area was a busy place. Here, four trucks dump at the same time supervised by the 'tip man'.

Figure 7.24 Taken in July 1952, this aerial photograph shows initial excavation work and pre-stripping of surface soil in progress. In the distance, a second 'cut' has just started. Notice the power line crossing the site from upper left to lower right, which will remain untouched throughout the operation.

168　CHAPTER 7　BRITISH OPENCAST COAL

Calow
September 1952

Figure 7.25 Taken in September 1952, only two months after the previous photograph, this view shows the rapid progress at Calow. The power line evident in the previous photo is seen at upper left.

CHAPTER 7 SIGNIFICANT SITES   169

Figure 7.26 This aerial shot taken in May 1953 shows the diverse and numerous fleets of machines employed by Tarslag at Calow.

Calow
May 1953

Figure 7.27 This view gives an idea of the extensive area encompassed by Calow Herne site. It was taken in March 1954, only 20 months after the first of these aerial photographs. Many more years' work would follow, with the land restored as new areas were opened up. External overburden tips and surface soil stockpiles can be seen.

Calow
March 1954

Figure 7.28 A rare machine on British opencast sites was this electric powered Ruston-Bucyrus 100-RB shovel of 3½ cubic capacity. It arrived at Calow a couple of years after start-up and could fill the Euclid 15-ton trucks in two or three passes.

Figure 7.29 Calow Herne site was not 'electrified', so Tarslag built this semi-portable, diesel-powered electric generator set with gravity-fed diesel tank to power the big 100-RB shovel. A long trailing cable connected to the shovel.

CHAPTER 7 SIGNIFICANT SITES 171

Figure 7.30 Hard rock at Calow is being drilled by two air-powered rock drills to prepare for blasting. The 100-RB shovel is excavating the rock, and a Ruston-Bucyrus 19-RB coal shovel is seen in the background.

Figure 7.31 The 100-RB shovel in difficulty after discovering some old underground mine workings. Notice the labourer with hand shovel working under the machine and the crew's lunch and tea mugs at lower right!

Figure 7.32 Bedford lorries wait their turn to be loaded by a Ruston-Bucyrus 19-RB shovel. Thin, steeply dipping coal seams are evident here, clearly demonstrating smaller machines work more efficiently than larger ones in these conditions.

Figure 7.33 For the extensive hard rock, Tarslag built this mobile compressor truck with three air compressors to serve three rock drills. The picture shows it being serviced by another Tarslag special, a service truck complete with diesel fuel, water, various oils, and a small compressor for tyres. Both vehicles are built on former 6 × 4 haul truck chassis.

CHAPTER 7 SIGNIFICANT SITES 173

Figure 7.34 Three new Euclid S-18 motor scrapers are proudly shown off by the foreman and drivers in 1957. The S-18 carried 18 cubic yards struck or 21 yards heaped, and power came from a single Detroit 6-110 diesel engine.

Figure 7.35 During the 1950s, several additional machines arrived at Calow, including this Ruston-Bucyrus 5-W walking dragline purchased new by Tarslag in 1954. It carried a five cubic yard bucket on a 120 foot boom. The dragline is exposing a steeply dipping seam, working from the spoil side. Further right the electric 100-RB shovel removes overburden while a coal drill prepares a lower bench for blasting.

Figure 7.36 No, they are not burying the 5-W dragline! Power lines are expensive items to remove, so it was cheaper to have scrapers dig this trench, allowing the dragline to walk under the power line as it moved from one section of the site to another.

Glyn Glas – South Wales

In 1967, the eyes of the opencast coal industry focused on Shand Mining when it made headlines by importing the first Marion 191-M electric shovel into the UK. The 15-yard excavator was assembled at the company's new Glyn Glas opencast site near Ammanford and, with an operating weight of 434 tons, was by far the biggest mining shovel yet seen in the UK. Built by Marion Power Shovel Company of Marion, Ohio, the 191-M had already gained an excellent reputation in the USA, when introduced in 1951 and was the world's largest shovel on two crawlers. The Glyn Glas machine was specified with a 50-foot eight-inch boom, 32-foot dipper handle and mounted on a heavy-duty undercarriage with 27-foot long crawlers and 42-inch wide shoes.

Matching the largest mining shovel in the UK to a suitably sized truck fleet resulted in a second record for Shand Mining at Glyn Glas. At that time, the largest hauler in production at the Euclid factory in Motherwell, Scotland was the 45-ton R-45. (The 65-ton R-65 was not yet available.) Aveling-Barford and Caterpillar could only offer trucks up to 35 tons capacity, so Shand imported six 75-ton Wabco 75A rear dump haulers to team with its new shovel, the largest yet to work in the UK.

Glyn Glas was an impressive site with interesting fleets of equipment, often changing over the years. The big Marion shovel was joined by a pair of Ruston-Bucyrus 120-RB five-yard electric shovels obtained second-hand from the National Coal Board. There was also a British-built three-yard Marion 101-M diesel shovel and the iconic seven-yard Lima 2400 dragline. Before the site finished, a Ruston-Bucyrus six-yard 150-RB electric shovel saw duty there, along with no less than three 3½-yard Ruston-Bucyrus 71-RB diesel shovels. In addition to the Wabcos, haul trucks keeping these shovels busy included 45-ton Euclid R-45s and 35-ton Aveling-Barford SN-35s.

A variety of relatively small excavators were employed to efficiently load the 'black diamonds' from steeply dipping and faulted seams. These included Ruston-Bucyrus ¾-yard 22-RB shovels, a 22-RB dragline, ¾-yard NCK 305 shovels and Hy-Mac 580 hydraulic excavators. There was also a 1½-yard NCK 605 shovel used for excavating coal or thick partings between coal seams. Coal was hauled to the nearest OE disposal point by various road-going lorries fitted with high-volume coal bodies.

Topsoil and subsoil salvage and replacement was handled by a fleet of Euclid TS-24 twin-engined motor scrapers. Caterpillar D9 tractors with blades and single-shank rippers push-loaded these scrapers and, for a while, were joined by a 440-horsepower Euclid 82-80 dozer powered by two GM 6-71N engines, and equipped with a cable-operated, inside-mounted narrow blade designed for push-loading duties.

The Shand contract at Glyn Glas ran for approximately ten years. It was followed in 1977 by the Glyn Glas South extension that was awarded to Shephard, Hill & Co. Ltd. A fresh set of equipment operated on this extension including a 12-yard P&H1900AL electric shovel. Coal production at Glyn Glas wound down in about 1982.

Figure 7.37 Assembly of the Marion 191-M shovel, Britain's largest mining shovel, takes place at Glyn Glas in 1967. *KH*

Figure 7.38 Six brand new 75-ton haul trucks arrive at Glyn Glas from America in 1967. The Wabco 75A trucks will work with the big Marion 191-M shovel. *KH*

176 CHAPTER 7 BRITISH OPENCAST COAL

Figure 7.39 Shown at work in 1970, the Marion 191-M with 15-yard dipper takes two passes to load a 45-ton Terex R-45 and three passes to load a 75-ton Wabco 75A. *KH*

Figure 7.40 The electric Marion 191-M loads a Wabco 75A hauler. The Shand machine was the first of five eventually imported into the UK from America. *KH*

CHAPTER 7 SIGNIFICANT SITES   **177**

Figure 7.41 Obtained from the National Coal Board, two older Ruston-Bucyrus 5-yard 120-RB shovels did their part at Glyn Glas. This one loads a 35-ton Aveling-Barford SN 35-ton dump truck. *KH*

Figure 7.42 This 120-RB electric shovel keeps a fleet of Euclid R-45 trucks on the move. Its ancestor is the Bucyrus-Erie 120-B of 1925, the very first quarry and mine shovel. *KH*

Figure 7.43 A Wabco 75A hauler dumps its 75-ton load at Glyn Glas. These haulers are powered by a 700-horsepower diesel engine driving through an Allison transmission with torque converter. *KH*

Figure 7.44 A variety of smaller shovels load coal into lorries with high-volume bodies. The NCK 605 shovel in the centre is excavating parting between coal seams. *KH*

CHAPTER 7 SIGNIFICANT SITES 179

Figure 7.45 A ¾-yard NCK 305 shovel excavates and loads coal into a high-volume bodied lorry for transport to the local off-site disposal point. Labourers remove contaminants to ensure a clean product; no hard hats necessary in 1968! *KH*

Figure 7.46 No thick, easily dug coal seams here! A 22-RB dragline, a 22-RB shovel, a pair of Hymac 580 hydraulic backhoes and a Caterpillar dozer work to separate partings from coal and load it cleanly into lorries. *KH*

Figure 7.47 Some idea of the complicated geology at Glyn Glas is revealed here. The thin, steeply dipping coal seam, measuring about 15 inches in thickness, is peeled off the face by the Hymac 580 backhoe on top, and loaded by the Ruston-Bucyrus 22-RB shovel. *KH*

Figure 7.48 A general pit view at Glyn Glas shows haul roads at different levels as the multiple seams are exposed.

CHAPTER 7 SIGNIFICANT SITES | 181

Figure 7.49 This Euclid 82-80 twin-engined crawler tractor fitted with cab and narrow cable-operated blade provides 440-horse pushing power for the Euclid TS-24 scrapers. *KH*

Figure 7.50 One of three Ruston-Bucyrus 71-RB 3½-yard shovels at Glyn Glas loads a Euclid R-45 dump truck. *KH*

Figure 7.51 Two of the 71-RB shovels excavate their way through difficult geological conditions at Glyn Glas. An extensive fleet of relatively small excavators is needed to cleanly extract the valuable anthracite. KH

Maesgwyn – South Wales

Over a period of 40 years, Maesgwyn opencast site and its several extensions yielded more than 11 million tons of top quality anthracite and literally moved a mountain in South Wales. Located high above the small town of Glynneath, Glamorganshire, the site commenced production in 1949, and was operated continuously by George Wimpey & Company Ltd. (later Wimpey Mining). The average mining ratio of 35 cubic yards of material moved to one ton of coal recovered was one of the highest in the world. Some coal seams had been previously mined by underground methods, raising the ratio even more. But the top quality anthracite, the efficient methods employed and the high coal recovery achieved made the whole proposition economically viable.

Prior to 1961, Maesgwyn's maximum coal output reached only 200,000 tons per year. One of the first machines to appear was a Marion 7200 walking dragline of six cubic yards capacity, but the relatively small machines available could only excavate to a limited depth. In 1955, when larger machines became available, Wimpey employed three six-yard Bucyrus-Erie 150-B electric shovels and several 34-ton capacity Euclid FFD tandem drive trucks imported from America. Wimpey also transferred in a 12-yard Marion 7400 dragline from a completed project in Australia.

A big boost came in 1961 when a Rapier W1800 walking dragline, owned by the OE, took its first bites at Maesgwyn. Weighing some 2,000 tons, and carrying a bucket holding 40 cubic yards on a 247-foot boom, the machine claimed, at that time, to be the largest dragline operating anywhere in the world. The W1800 worked low down in the huge excavation, uncovering the lowest coal seams. Overburden above the dragline, ranging up to 430 feet deep and including more coal seams, was removed in a series of 40-foot benches by fleets of shovels and dump trucks.

In 1963, the shovel fleet consisted of three electric Bucyrus-Erie 150-Bs, a Ruston-Bucyrus 150-RB and a diesel Lima 2400, all of six-yard capacity. The tough, rocky overburden was drilled by Ingersoll-Rand DM3 and up to five Bucyrus-Erie 40-R drills to prepare for blasting. The 40-R drills put down 6¼-inch holes up to 65 feet deep, at 15–21-feet centres. The holes were loaded with ammonium nitrate/fuel oil and fired with gelignite primers and multi-second delay detonators.

Haul trucks comprised mostly of Euclid R-27s and R-45s with a few Aveling-Barford SN 35-ton and AEC 30 ton for variety. The 45-ton Euclid R-45s were some of the very first used in the UK. In other areas of the site, where overburden depths were not as great, a five-yard Ruston-Bucyrus 110-RB crawler dragline and the Marion 7400 maintained production. Three 38-RB shovels and two 22-RB draglines looked after coal cleaning and loading; coal was hauled to the local OE disposal point by some of Wimpey's extensive fleet of AEC Mammoth Major highway lorries of ten and 14 cubic yards capacity.

During 1964 and 1965, Wimpey added three Bucyrus-Erie 190-B shovels to its fleet. Purchase of these American nine-yard electric shovels, the first in the UK, proved again that Wimpey and the Maesgwyn site were at the cutting edge of excavation technology, as this class of shovel was not made available by Ruston-Bucyrus at Lincoln until about ten years later. They replaced the three 150-B shovels, but a new 150-RB was added, maintaining a fleet of six shovels including the older 150-RB and Lima 2400. At the same time, the hauler fleet was upgraded to 26 Euclid R-45s. All other haul trucks were retired, but two more 38-RB coal shovels were added.

Wimpey continued to update its equipment. By 1967, the Euclid R-45 fleet extended to 30 units, and the Lima 2400 was replaced with a new one. The diesel-powered Lima was essential to the site as it could move to any location for short-term jobs without the hindrance of a power cable and substation. Starting in the early 1970s, Maesgwyn again revamped the site with modern equipment. Older machines were retired and replaced with new ones of larger capacity. These included P&H 1900AL and Ruston-Bucyrus 195-B electric shovels in the 12-yard class. In 1976, Wimpey erected at Maesgwyn one of five 17-yard Marion 195-M crawler draglines it imported from America.

The site ran its course until the late 1980s. In 1988, the Rapier W1800 was sold to Beltrami Enterprises in Pennsylvania, USA, but never worked again. Maesgwyn was restored progressively over the decades, with most of the site being returned to the Forestry Commission, which planted conifer seedlings on the graded slopes. These thrived on the clean, weed-free shaley soils; what was previously poor scrubland is now covered with pine trees. The land that once yielded its valuable resource is now returned to nature

Figure 7.52 Marion 7400 with 175-foot boom and 12-yard bucket worked about 20 years at Maesgwyn after being transferred from Australia by Wimpey. It was then sold to Alberta, Canada. *KH*

CHAPTER 7 BRITISH OPENCAST COAL

Figure 7.53 One of three Bucyrus-Erie 150-B shovels working in 1963, imported before the 150-RB was built by Ruston-Bucyrus at Lincoln. *KH*

Figure 7.54 The six-yard 150-RB electric shovel loads an Aveling-Barford SN dump truck. *KH*

CHAPTER 7 SIGNIFICANT SITES | 185

Figure 7.55
Maesgwyn's busy dump area in 1963 with Euclid R-22, R-27, and R-45 trucks keeping the 'tip man' busy. *KH*

Figure 7.56 Aveling-Barford SN 35-ton haulers augmented the fleet of predominantly Euclid models. *KH*

186 CHAPTER 7 BRITISH OPENCAST COAL

Figure 7.57 This Rapier W1800 dragline with 247-foot boom and 40-yard bucket was the world's largest when commissioned in 1961. *KH*

Figure 7.58 The Rapier W1800 dumps another 40 yards of blasted rock to uncover the lower seam.

CHAPTER 7   SIGNIFICANT SITES   **187**

Figure 7.59 An International Drott crawler loader and 22-RB shovel work with coal exposed by the giant Rapier W1800 in the background. *KH*

Figure 7.60 A Lima 2400 loads a Euclid 45-ton hauler. The diesel-powered 6-yard shovel enjoyed superior mobility being free from a power cable.

188 **CHAPTER 7** BRITISH OPENCAST COAL

Figure 7.61 A pair of Bucyrus-Erie 40-R blast hole drills prepare the rocky overburden for the shovels. Well-blasted material increases shovel availability and reduces repair costs. *KH*

Figure 7.62 This lower seam at Maesgwyn averaged 10 feet thick, but here two 38-RBs excavate a geological irregularity where the seam thickens to 25 feet of solid anthracite! *KH*

CHAPTER 7 SIGNIFICANT SITES 189

Figure 7.63 Erection of the second Bucyrus-Erie 190-B is almost complete. Three of these 9-yard shovels appeared at Maesgwyn beginning in 1964. *KH*

Figure 7.64 The third 190-B shovel at Maesgwyn loads a 45-ton Euclid hauler. This particular hauler was one of the first R-45s built in Scotland and had been at Maesgwyn for two years when this picture was taken in 1965. *KH*

190 CHAPTER 7 BRITISH OPENCAST COAL

Figure 7.65 Three 9-yard Bucyrus-Erie 190-Bs make short work of a 40-foot bench, and a pair of Bucyrus-Erie 40-R drills puts down holes for the next blast.

Figure 7.66 In this 1972 shot, a late version AEC 8 x 4 lorry receives a load from a new Ruston-Bucyrus 38-RB shovel. The banksman keeps the machine clean as it works, a practice that would not be allowed today!

CHAPTER 7 SIGNIFICANT SITES 191

Figure 7.67
Maesgwyn's Marion 7400 removes overburden while a 22-RB dragline helps to clean out old underground workings in preparation for coal loading.

Figure 7.68 Employing road-going lorries to haul coal from site to disposal point was commonplace in the UK, and still is. This 38-RB shovel loads a late version AEC 8 x 4 lorry while a 22-RB dragline can be seen working against the high wall.

192 CHAPTER 7 BRITISH OPENCAST COAL

Figure 7.69 The first 12-yard P&H 1900AL shovel from Harnischfeger Corporation in America was tested at Maesgwyn in 1976; it's shown loading a 50-ton Terex R-50 hauler. Four more 1900AL's arrived in the UK soon afterwards.

Figure 7.70 The 40 cubic yard bucket of the Rapier W1800 dwarfs a 38-RB 1 ½-yard coal shovel. *KH*

Newman Spinney – Derbyshire

Newman Spinney Restoration Site, located between the village of Spinkhill and the M1 motorway in north-east Derbyshire, was an opencast coal site with a difference. It resulted from a failed experiment carried out jointly by the Central Electricity Generating Board and the East Midlands Electricity Board to gasify coal underground by applying heat. The gas project ended, but the coal seams were still on fire, emitting toxic gases and endangering two nearby underground mines. No one knew to what extent the coal had burned, so the only solution was to excavate the entire affected area, remove any remaining coal and extinguish the fire in the process.

In 1962, Murphy Brothers Ltd. of Syston, Leicestershire was awarded a contract to move 26 million cubic yards of overburden and retrieve any remaining coal by opencast methods. Since the amount of recoverable coal was undetermined, the contractor was paid to move the overburden at a price per cubic yard, and any coal found could be sold as a bonus. As the job progressed, it was found that relatively little coal had been burned, and by the time all was extracted in 1966, Murphy had recovered well over one million tons of coal!

Murphy employed an interesting variety of earthmoving machines comprising various makes of shovels, draglines and scrapers. At the start of the job, a new Ruston-Bucyrus 150-RB electric shovel with six-yard dipper was purchased, and as overburden depth increased, more equipment was added. A 2½-yard Ruston-Bucyrus 54-RB shovel and two aging 100-RB electric shovels with 3½-yard dippers were brought in. During the later phases of the contract, a third 3½-yard shovel, a Ruston-Bucyrus 71-RB, joined the fleet.

A mixed fleet of five draglines also assisted in overburden removal throughout the Newman Spinney contract. These were a Lima 2400 with seven-yard bucket, a Ruston-Bucyrus 5-W walker equipped with five-yard bucket, a Marion 111-M with 4½-yard bucket and two 2½-yard Ruston-Bucyrus 54-RBs. The draglines mostly uncovered the lower seams of coal, working in the bottom of the pit where they could dump directly into the adjacent worked-out area. Sometimes they worked in deeper areas pulling back spoil to fully expose a coal seam, often working in tandem and passing material from one to the other.

Each excavated cut followed the natural dip of the multiple seams, running from the shallow 'outcrop' end to the deepest part at the site boundary. The cuts grew progressively deeper over the course of the contract, culminating in a final high wall some 350 feet in height, one of the deepest achieved in the Midlands. The shovels loaded a fleet of 22- to 27-ton capacity dump trucks, mostly Euclid R-22, R-24 and R-27 models. The trucks hauled the material up the sloping benches to the shallow part of the site, then across to the opposite side of the excavation to backfill the pit where the coal had been removed.

A resident fleet of Caterpillar motor scrapers (14 DW21s and 6 631s) assisted by Caterpillar D8 and D9 bulldozers performed their usual duty of salvaging and replacing topsoil and subsoil, but they also played a significant role in overburden removal. They built roads for the haul trucks and removed up to ten feet of interburden material between coal seams. The open-bowl scrapers were also sometimes used as haulage vehicles, loaded by the shovels, when dump trucks were at a premium due to long hauls.

Coal seams were loaded by a fleet of five Ruston-Bucyrus 22-RB cable shovels, some equipped with oversize coal dippers. The coal was so plentiful that even the 2½-yard 54-RB shovel excavated some coal, an unusual application for such a large machine in the early 1960s. The shovels loaded a fleet of Leyland Hippo 6 × 4 heavy duty lorries carrying about 15 tons per load from the pit to an on-site coal washing plant. From there, road-going lorries transported the washed coal to local power stations.

Towards the end of 1966, the last coal was loaded out when the 350-foot final high was reached. Then followed a massive year-long project to backfill millions of cubic yards of stockpiled spoil into the huge void and bring it up to approximate original elevation. Some shovels worked on top of the spoil pile to load the trucks that hauled the material. The scraper fleet was augmented by three Caterpillar 657s, push-loaded by up to four crawler tractors, to contour the final landscape.

By the end of 1967, site restoration was complete, including installation of land drains and boundary fences. Today, travellers on the M1 motorway, about a mile north of Junction 30, observe only green agricultural land showing no sign of the massive excavation that took place there more than 50 years ago.

Figure 7.71 Murphy Brothers purchased a brand new Ruston-Bucyrus six-yard 150-RB electric shovel for the five-year Newman Spinney contract. *KH*

Figure 7.72 Here the 150-RB electric shovel loads a Euclid R-27 hauler, while an Aveling-Barford grader keeps the area clear of spilled material. *KH*

CHAPTER 7 SIGNIFICANT SITES **195**

Figure 7.73 One of the many draglines at Newman Spinney was this diesel-powered 4½-yard Marion 111-M. *KH*

Figure 7.74 Some idea of the tough rock encountered at Newman Spinney is seen here as one of two Ruston-Bucyrus 3½-yard 100-RB electric shovels loads a 22-ton British-built Euclid R-22 dump truck. *KH*

Figure 7.75 A variety of draglines worked at Newman Spinney including this 4½-yard Marion 111-M and seven-yard Lima 2400 on the lower bench. *KH*

Figure 7.76 This British-built diesel-powered 5-W walking dragline swings a five-yard bucket on a 135-foot boom. *KH*

198 CHAPTER 7 BRITISH OPENCAST COAL

Figure 7.77 An action shot of a Euclid R-22 truck speeding from an upper overburden bench to the tip. *KH*

Figure 7.78 One of Murphy's 27-ton Euclid R-27, starting its trip to the tip, passes the 5-W dragline as it receives fuel from the tanker lorry. *KH*

CHAPTER 7 SIGNIFICANT SITES 199

Figure 7.79 A substantial amount of overburden at Newman Spinney was removed by scrapers. Here Caterpillar DW-21s are push-loaded, three at a time, by Caterpillar D8 bulldozers. *KH*

Figure 7.80 A Caterpillar DW-21 scraper is push-loaded by a Caterpillar tractor equipped with a cable-operated blade reinforced in the centre for this heavy push-loading duty. *KH*

Figure 7.81 Two draglines work in tandem: a Ruston-Bucyrus 54-RB pulls back spoil from the coal, while a Lima 2400 uncovers one of the lower coal seams at Newman Spinney. *KH*

Figure 7.82 Three draglines work as a team to pull back spoil. The lower 5-W passes material to a 54-RB, which in turn passes it to a Marion 111-M. *KH*

CHAPTER 7 SIGNIFICANT SITES 201

Figure 7.83 Two Ruston-Bucyrus 22-RB shovels load coal into Leyland Hippo lorries equipped with special high-volume coal bodies. *KH*

Figure 7.84 A large coal-loader for the early 1960s, this 2½-yard 54-RB shovel loads coal directly from the seam into a Leyland Hippo lorry. *KH*

Figure 7.85 Digging deeper at Newman Spinney, a modern 3½-yard Ruston-Bucyrus 71-RB shovel joins the fleet. *KH*

Figure 7.86 In 1967, Caterpillar twin-engined 657 scrapers of 44-yard capacity joined the resident scraper fleet for the massive job of filling in the final void. *KH*

Figure 7.87 The final high wall reached some 350 feet in height, one of the highest in the Midlands. The 150-RB shovel completes overburden removal at the lower level while a 22-RB loads coal. *KH*

Ox-Bow – Yorkshire

One of the oldest established opencast areas in the UK was at Templenewsam south east of Leeds, Yorkshire. From its beginnings in the early 1940s, Sir Lindsay Parkinson & Company Ltd. (Fairclough Parkinson Mining after 1974) worked a succession of contracts spanning almost five decades. One of the longest-running was Ox-Bow and its extensions Charcoal, Gamblethorpe and Skelton.

Parkinson commenced Ox-Bow operations in 1957 with a Rapier W600 walking dragline, one of only four of this type built by the company, and fitted with an 11-yard bucket on a 186-foot boom. Other equipment arriving at the start included a new Ruston-Bucyrus 110-RB 4½-yard shovel, the first of this type to work in UK opencast coal, and an old five-yard Ruston-Bucyrus 5-W walking dragline from an adjacent site.

Ox-Bow was one of the very few sites in the UK to employ a crane hoist for raising coal from the deepest cuts, thereby eliminating construction of costly ramps for lorries. For Ox-Bow, Parkinson did not erect the elaborate rail-mounted derrick system adopted at Acorn Bank already described. Instead, a Rapier W150 walking dragline was converted to a crane hoist. Coal shovels loaded a large skip, which was then hoisted by the crane and dumped into a portable hopper from which road lorries were filled. A variety of ⅝-yard and ¾-yard shovels loaded coal, mostly Ruston-Bucyrus 19-RB, 22-RB, and NCK 304 models. Another claim to fame for Ox-Bow was the arrival in 1958 of the very first Ruston-Bucyrus 150-RB electric shovel, a machine that would become very popular (and frequently mentioned in this book!). The 237-ton shovel carried a six-yard dipper on a 37-foot six-inch boom.

After Ox-Bow wound down, Parkinson continued its opencast work in the same area, as the company succeeded in obtaining three more contracts, all running more or less concurrently. These new sites were Charcoal, Gamblethorpe and first phase of Skelton, and afforded men and machines continual employment up to 1992. Coal from these sites was hauled to Templenewsam disposal point which Parkinson had operated since the early 1940s. From there, most of the screened coal was hauled to nearby Skelton Grange Power Station.

Figure 7.88 The main excavator at Ox-Bow was this Rapier W600 walking dragline, one of only four built by the company. The machine remained active in the Templenewsam area for its entire life until scrapped about 1995. *KH*

CHAPTER 7 SIGNIFICANT SITES   205

Figure 7.89 The Rapier W600 looms very large for a dragline carrying a bucket of only 11 cubic yards on a 186-foot boom. *KH*

Figure 7.90 View of Ox-Bow showing the Rapier W150 crane hoist for raising coal and filling lorries via the portable hopper. In the foreground a 5-W walking dragline on the spoil side pulls back material dumped by the W600 dragline (not in the picture). In the background, a parked 150-RB and several 22-RB shovels work in the coal. *KH*

Figure 7.91 A pair of Ruston-Bucyrus 22-RB shovels load the coal skip, which will be hoisted and dumped into the hopper by the Rapier W150 crane hoist. The portable hopper can be seen at left. *KH*

CHAPTER 7 SIGNIFICANT SITES **207**

Figure 7.92 Overview of upper coal seams being loaded by Ruston-Bucyrus 19-RB, 22-RB and NCK 304 shovels. The site contained several good-quality coal seams. *KH*

Figure 7.93 This Ruston-Bucyrus 110-RB shovel arrived new for the start of the Ox-Bow contract in 1957. Here it loads a British-built Euclid R-22 dump truck. *KH*

Figure 7.94 The 4½-yard Ruston-Bucyrus 110-RB electric shovel at Ox-Bow was one of the first to leave the Lincoln factory. *KH*

CHAPTER 7 SIGNIFICANT SITES 209

Figure 7.95 A fleet of new Aveling-Barford SN 35-ton haul trucks arrived at Ox-Bow in the mid-1960s. This one is being loaded by the 110-RB shovel. *KH*

Figure 7.96 One of Parkinson's company-wide fleet of Ruston-Bucyrus 5-W walking draglines loads a 35-ton Aveling-Barford SN haul truck. Built in 1946 the five-yard machine was employed at Ox-Bow until 1964 when it was scrapped. *KH*

Figure 7.97 This special coal-loading dipper on a 22-RB shovel is most efficient at loading thin coal seams. *KH*

CHAPTER 7 SIGNIFICANT SITES 211

Figure 7.98 In April 1962, Ox-Bow celebrated the millionth ton of coal extracted from the site.

Figure 7.99 Activity at Ox-Bow with no mistake who owns the machines! Two Ruston-Bucyrus 22-RB shovels load coal in the foreground, as a Ruston-Bucyrus 150-RB shovel in the centre moves to a new location, led by a Caterpillar D8 bulldozer pulling a pair of skid-mounted cable stands, affectionately known as 'goal posts'. Behind these the Rapier W600 towers over the pit, while in the background a blast hole drill generates plenty of dust and the Rapier W150 crane hoist and hopper complete the scene. *KH*

Figure 7.100 For a short time, a horizontal coal auger was used to gain some extra coal tonnage from the edge of the site beyond excavation limits. Auger methods of this type usually produced about 50 per cent coal recovery from holes drilled to a depth of about 70 feet. Despite a low recovery, it is coal that would otherwise never be recovered. *KH*

Figure 7.101 The original 5-W dragline at Ox-Bow, built in 1946, was scrapped in 1964. At the same time it was replaced by a slightly newer 5-W seen here under erection. *KH*

Figure 7.102 An Aveling-Barford SN35 dump deposits its load at the tip. The 35-ton Rolls-Royce powered truck was built at the company's factory at Grantham. KH

Pen Gosto – South Wales

A large opencast site operated by Sir Lindsay Parkinson & Company Ltd. was Pen Gosto, located near Gwaun-cae-Gurwen at the western edge of the Black Mountains. The site started in 1968 and ran until the end of the 1970s, and called for extraction of 2.2 million tons of anthracite from multiple seams with a high overburden-to-coal ratio. The site was known for a variety of different excavating machines, which changed a number of times throughout its duration. At the start, Parkinson moved in two walking draglines from its fleet: a diesel-powered 1949 Marion 7200, and an electric-powered Marion 7400, originally purchased in 1957 for a site at Kippax, Yorkshire.

The 7200 was typically specified with a seven-yard bucket on a 120-foot boom. It was powered by a single Dorman engine to drive the hoist and drag drums by friction clutches and brakes, and also to power the single electric swing motor through a belt-driven generator. Overall machine weight in operation was 237 tons; its circular tub was 24 feet in diameter. The Marion 7400 was equipped with a 215-foot boom and carried a 12-yard bucket. An AC motor-generator set in the rear of the machine received power from a trailing cable and provided DC power to the main hoist/drag motor and two swing motors.

As well as the two draglines, Pen Gosto commenced with two electric shovels to assist with overburden removal: a Ruston-Bucyrus 150-RB with six-yard dipper and a new Marion 191-M with 12-yard dipper. The 191-M, one of only five imported into the UK, was the largest two-crawler mining shovel to work in the UK at that time. Normally the 191-M of that era was specified with a standard 15-yard dipper, but no doubt because of the hard rocky overburden at Pen Gosto, Parkinson chose a 12-yard heavy-duty dipper to prolong the life of the machine. A fleet of 45-ton capacity Terex R-45 rear dump trucks served the 150-RB shovel, while the larger 191-M shovel kept a fleet of 65-ton Terex R-65 trucks busy. To better match the capacity of the big Marion 191-M, Parkinson purchased four 100-ton Unit Rig M100 Lectrahaul dump trucks. These diesel-electric vehicles were powered by 1,000-horsepower Caterpillar engines and GE electric motors in each rear wheel.

By 1972, a second 150-RB and a 71-RB shovel appeared, and the following year a third walking dragline, an electrically powered Rapier W300, started work. It was one of two originally purchased by Parkinson in 1957, and moved from a site in County Durham. The W300 had a seven-yard bucket on a 140-foot boom and, like the Marion 7200 and 7400 draglines, utilized a single motor to power the hoist and drag drums through clutches and brakes.

In 1977, a new Marion 191-M with hydraulic propel, the fifth imported into the UK, was erected to replace the first, which was dismantled and sold. A third Marion 191-M also appeared at Pen Gosto when Parkinson moved it from Dunraven, having completed its work there. At the same time, a 12-yard Ruston-Bucyrus 195-B was added to the fleet. A second batch of Unit Rig M100 trucks arrived to work with the new shovels.

Other haulers appearing over the years included Aveling-Barford Centaur 40s, Terex R-50s and 85-ton Terex 33-11s. Overburden drills included up to three Haus Herr rigs and a Bucyrus-Erie 30-R.

Coal loading was initially handled by two Ruston-Bucyrus 30-RB cable excavators, soon followed by a hydraulic Hymac 1290 with coal dipper. These loaded a fleet of AEC 15-ton capacity dump trucks that travelled on private haul roads to the disposal point. Later, some of the Terex R-45s were assigned to coal duty and a Caterpillar 988 wheel loader joined the fleet. By 1977, the cable excavators had been replaced with Hymac 890s.

Figure 7.103 Parkinson's Ruston-Bucyrus 150-RB loads another six yards of overburden onto a Terex R-45 haul truck. *KH*

CHAPTER 7 SIGNIFICANT SITES 215

Figure 7.104 A unique combination indeed! The seven-yard Rapier W300 dragline loads an electric-drive 100-ton capacity Unit Rig M100 haul truck with surface soil. *KH*

Figure 7.105 No less than three Marion 191-M electric shovels saw service at Pen Gosto. This first machine, fitted with a 12-yard dipper, worked approximately ten years before being replaced with another new 191-M, more modern and with hydraulic propel. The 191-Ms were the largest two-crawler mining shovels to work in the UK at that time. *KH*

Figure 7.106 One of Parkinson's two Marion 7400 draglines working at Pen Gosto to remove overburden with a 12-yard bucket on a 215-foot boom. *KH*

Figure 7.107 The Marion 7200 diesel-driven dragline swings a seven-yard bucket on a 120-foot boom. *KH*

CHAPTER 7 SIGNIFICANT SITES **217**

Figure 7.108 The 1949 vintage Marion 7200 dragline undergoes some maintenance and repair. *KH*

Figure 7.109 Coaling operations at Pen Gosto show a pair of Ruston-Bucyrus 30-RB shovels loading 15-ton AEC lorries with high-volume coal bodies. The Marion 7400 dragline in the background removes overburden. *KH*

218 CHAPTER 7 BRITISH OPENCAST COAL

Figure 7.110 A busy coal-loading scene with an International Drott crawler loader, Caterpillar 988 wheel loader and Ruston-Bucyrus 30-RB shovel. *KH*

Figure 7.111 This overall view shows the Marion 7200 excavating the lower level to the right, the Marion 191-M shovel forming the working bench at centre, and the 150-RB shovel taking the upper bench at left. In the distance, the Rapier W300 dragline is engaged in removing surface soil and loading M100 Lectrahaul trucks. *KH*

Radar North Group – Northumberland

In 1955 the OE purchased two new Marion 7800 walking draglines for use by James Miller & Partners at Radar South, first of a long succession of opencast coal sites in the Widdrington area. Originally purchased with 22-yard buckets and booms 280 feet long, the 1,580-ton machines were ranked as some of the largest operating anywhere. The new Marions started work in late 1956, and the OE organized a celebration to launch the biggest digging machines ever seen in Britain.

After working only six weeks, disaster struck! In February 1957, the boom on one of the draglines collapsed. No one was injured, but the accident resulted in protracted investigations lasting several months. The second dragline was taken out of commission during this time until the issues were resolved. Apparently before the accident, it was reported that the long booms appeared too highly stressed and showed alarming whip. It was also revealed that daily boom inspections which should have been normal practice, were not carried out. When the booms were finally replaced, they were reduced in length to 240 feet, and the draglines operated successfully with 30-yard buckets for the rest of their working days.

Following completion of Radar South, both 7800s moved to the adjacent and much larger Radar North site under contract to Derek Crouch. Among several machines unique to the British opencast scene at Radar, none drew more attention than the Krupp bucket wheel excavator, the only such machine to operate on a UK opencast coal site. Use of a bucket wheel excavator requires reasonably level topography, long straight cuts and fairly soft material free of boulders. To take advantage of its high-capacity continuous output, and after geological testing proved conditions were favourable, Derek Crouch ordered the machine from Krupp Industrietechnik of Germany. Erected in 1960, the 340-ton machine carried eight buckets of ⅓-cubic yard capacity and could dig a bank height of 40 feet at a design rate of 1,200 cubic yards per hour.

The site was laid out with the wheel excavator taking the upper 40 feet of glacial drift. It moved back and forth from end to end as it excavated the working face. It discharged material onto a mobile conveyor bridge that, in turn, discharged through a rail-mounted hopper car onto the main conveyor running the length of the cut.

From his cab on the wheel boom, the wheel operator controlled the boom hoist, boom swing, wheel rotation and machine propel. Excavation sequence began at the top of the face with the wheel rotating into the bank. The boom was slowly swung from side to side, lowering a couple of feet at each pass until the bottom of the face was reached. The excavator then moved forward and the sequence repeated. A second operator, positioned in a cab on the discharge boom, controlled that boom's swing and also movement of the discharge hopper. When the excavator moved over to the next cut, the main conveyor was skidded laterally by crawler tractors to the new position.

From the main conveyor, spoil material transferred to a second conveyor running around the end of the cut, then transferred again to a third conveyor running onto the area being filled and reclaimed. This conveyor ran the full length of the fill, roughly parallel to the cut, and a rail-mounted 'tripper car' collected the material from the belt and delivered it to the spreader. The tripper was simply a section of inclined conveyor where the belt doubled back on itself, allowing the material to be dumped onto a 60-foot long feeder conveyor, and finally onto a 100-foot long discharge conveyor. The spreader was controlled by an operator who placed material in an orderly fashion onto the spoil piles. Since this material made up the final layer being placed prior to subsoil and topsoil, correct placement by the operator greatly minimized levelling costs that otherwise would be done by bulldozers. The two operators on the wheel excavator and the operator on the spreader were necessarily in radio contact at all times. On any particular shift, possibly to minimize boredom, the operators interchanged their positions, spending no more than a few hours at one station. Well over a mile of conveyor installation was involved with this elaborate bucket wheel system.

Below the wheel excavator, the next 80 feet of overburden, consisting of hard rocky material, was removed in two lifts by shovels and trucks to reach the first coal seam. Initially a six-yard Bucyrus-Erie 150-B shovel was employed to load a fleet of Caterpillar PR-21 rear dump wagons. Based on the successful

DW-21 motor scraper, introduced in 1951, these haul units consisted of the DW tractor of 300-horsepower coupled to an Athey rear-dump wagon of 34 tons capacity. Up to 14 of these units were used at Radar, but by the mid-1960s, they were retired and replaced by Aveling-Barford SN-35 and Caterpillar 769 rear dump trucks of 35 tons capacity.

Positioned on a wide, level bench formed after coal had been removed from the upper seam, the Marion 7800 draglines dug down a further 60–90 feet to uncover the lowest coal seam and deposited excavated material into the adjacent mined-out pit.

A large fleet of scrapers, bulldozers and motor graders were all kept busy removing salvageable soil and assisted in overburden removal by reducing the depth of dig for the shovels. During the 1960s the fleet included Euclid TS-14 and Caterpillar 657 twin-engined scrapers, along with a 34-ton Russian dozer Model D572 of 300-horsepower. Wheel loaders also played their part in soil handling when haul distances were more than half a mile. This fleet included Caterpillar 988 (six-yard) and 992A (ten-yard) loaders teamed with Caterpillar 35-ton 769 dump trucks.

As overburden depths increased in the late 1960s, shovels and trucks of larger capacity became necessary. The shovel fleet was augmented by the addition of a seven-yard 150-RB and a ten-yard Marion 182-M. Correspondingly larger haul trucks arrived in the form of five 50-ton Caterpillar 773 and four 100-ton diesel-electric Unit Rig M-85s. A fleet of 65-ton Michigan T-65 haul trucks was tested by Clark Equipment's Michigan Division and announced as the first of a new range of haulers from the company. Wimpey and Costain also tested the units, and plans for manufacture in the UK were advanced, but after these tests Clark decided to cease production of haulers and the fleet was exported to continental Europe.

Haul road maintenance and loading area clean-up were handled by Aveling-Barford 99-H and Caterpillar 14E graders, and Michigan 180 and Caterpillar 824 wheel dozers. Excavators assigned to coal loading included Ruston-Bucyrus 22-RB, 30-RB, NCK 605 and Priestman Lion, supplemented in later years by Michigan, Caterpillar and Weatherill wheel loaders. Coal was transported from the pit to the local coal disposal point by AEC 8 × 4 road lorries and about 20 15-ton AEC 690 dump trucks. These were fitted with high-volume 'coal' bodies.

In 1969 Radar made headlines again when the largest dragline in Europe was commissioned. Owned by Derek Crouch and named 'Big Geordie', the Bucyrus-Erie 1550-W was imported in components from America and erected over a 12-month period. It boasted a 65-yard bucket, a 265-foot boom and an operating weight of 3,200 tons. Overall width was 77 feet eight inches, and each shoe measured 56 feet by ten feet. It replaced one of the Marion 7800 draglines, which was sold to a mine in Alberta, Canada.

The Radar North site continued well into the 1970s, and by that time Derek Crouch had been awarded contracts for the neighbouring Coldrife, Ladyburn, Radcliffe and Acklington opencast sites in the same area of Northumberland. These were connected by internal haul roads and the area became known as the Radar North Zone. Often machines would move from one site to another as the need arose. Radcliffe and Ladyburn sites were each home to one of the 25-yard Bucyrus-Erie 1150-B draglines moved over from the former Acorn Bank site.

In 1970, the 1550-W 'Big Geordie' moved from Radar North to the adjacent Sisters site, also operated by Derek Crouch. In 1977, she was replaced at Sisters by one of the new Bucyrus 1260-Ws in 1977, and Big Geordie set off across country on a seven-mile, specially built walk road to the newly opened Butterwell site. The 1550-W Big Geordie was the main excavator here, hired by Crouch to Taylor Woodrow, who won the contract. Butterwell eventually yielded approximately 12 million tons of coal.

CHAPTER 7 SIGNIFICANT SITES 221

Figure 7.112 This significant historical photograph was taken in 1956 during the raising of the booms on the two Marion 7800 draglines at Radar South. Apparently, both draglines were built simultaneously, and both booms were raised at the same time.

Figure 7.113 Dignitaries were invited to a ceremony in 1956 to launch the Marion 7800s, the largest digging machines ever seen in Britain.

222 CHAPTER 7 BRITISH OPENCAST COAL

Figure 7.114 Disaster struck only six weeks after the new draglines went to work. The 280-foot boom on one of them collapsed. After a lengthy delay, both were replaced with shorter 240-foot booms. *Peter Grimshaw collection*

Figure 7.115 Derek Crouch purchased the Krupp wheel excavator in 1960 for its Radar North contract. The upper boom supported and swung with the discharge boom, while the wheel boom swung independently. Specified output was 1200 cubic yards per hour. *KH*

CHAPTER 7 SIGNIFICANT SITES 223

Figure 7.116 The digging end of the bucket wheel excavator shows the wheel, which swings from side to side as it excavates the face from top to bottom. The pair of wide crawler tracks keep ground pressure to a manageable limit. *KH*

Figure 7.117 The Krupp bucket wheel excavator discharges into the rail-mounted hopper, which travels along the main conveyor. This is positioned parallel to the working face and as the excavator moves on its crawlers, the hopper car moves with it, running on rails that are an integral part of the conveyor assembly. A second operator in the cab on the discharge boom controls boom movement and hopper travel. *KH*

Figure 7.118 This crawler-mounted spreader, with 100-foot boom receives spoil material from the feed conveyor and 'tripper', which collects material from the main conveyor running the full length of the fill. *KH*

Figure 7.119 Working in 40-foot benches, shovels remove the harder rocky overburden below the level of the bucket wheel excavator. Here a Bucyrus-Erie 150-B six-yard shovel loads a 35-ton Aveling-Barford SN-35 haul truck. *KH*

Figure 7.120 Up to 14 Caterpillar PR-21 rear-dump haulers of 34 tons capacity made up the original overburden haulage fleet. *KH*

Figure 7.121 This view shows the three main stages of overburden removal at Radar North. At top left, the bucket wheel excavator takes the alluvial material. Below this, a Bucyrus-Erie 150-B electric shovel loads harder overburden into trucks; the Marion 7800 dragline exposes the lowest coal seam. *KH*

226 CHAPTER 7 BRITISH OPENCAST COAL

Figure 7.122 A Rapier W150 walking dragline with a six-yard bucket works at the side of the main cut to uncover shallow coal. Later, Crouch moved its other W150 from Abercrave in South Wales to the Coldrife site, adjacent to Radar North. *KH*

Figure 7.123 Caterpillar 657 and Euclid TS-14 scrapers push-loaded by Caterpillar D8 dozers move and level surface soil. In the distance a Caterpillar 992 wheel loader loads a subsoil pile into a fleet of Caterpillar 769 haulers for a long haul. *KH*

CHAPTER 7   SIGNIFICANT SITES   **227**

Figure 7.124 On hire from UMO Plant Ltd., this Russian-built 34-ton D572 bulldozer joined the soil handling fleet for a while. *KH*

Figure 7.125 A new seven-yard 150-RB complemented the shovel fleet in 1971. Here it loads a fleet of 50-ton Caterpillar 773 haulers. *KH*

Figure 7.126 Also arriving in 1971, this ten-yard Marion 182-M shovel and four Unit-Rig M85 100-ton haulers boosted overburden productivity. *KH*

Figure 7.127 Four of these Unit-Rig M85 haulers of 100 tons capacity worked with the Marion 182-M shovel. *KH*

Figure 7.128 These 65-ton Michigan T65 haulers were tested at Radar North, and at least two other coal sites in the UK. But after a short production run, Clark Michigan pulled out of the hauler business. *KH*

Figure 7.129 'Big Geordie', a Bucyrus-Erie 1550-W, began digging at Radar North in 1969. With 265-foot boom and 65-yard bucket, it claimed to be the largest dragline in Europe. *KH*

Figure 7.130 Twin hoist ropes and twin drag ropes manipulate the 65-yard bucket on 'Big Geordie'. Owner Derek Crouch (Contractors) Ltd. employed the 3,200-ton machine at Radar North followed by the Sisters site until 1977. She was then moved to the Butterwell site on hire to Taylor-Woodrow.

Figure 7.131 The two cabs of the 1550-W allowed the operator to select the best view, depending on which side he was dumping. Two Caterpillar D8 dozers are pushing back the 'roll' in front of the machine. *KH*

CHAPTER 7 SIGNIFICANT SITES **231**

Figure 7.132 A Ruston-Bucyrus 30-RB and Michigan wheel loader load coal into an AEC 690 dump truck. *KH*

Figure 7.133 A busy coal-loading scene at Radar North shows Caterpillar 930 wheel loaders, AEC dump trucks and a Caterpillar 824 wheel dozer cleaning the surface. *KH*

Figure 7.134 Crouch operated several other sites in the same area as Radar North, all connected by internal haul roads. Coldrife and Radcliffe sites each employed one of the OE's Bucyrus-Erie 1150-B 25-yard draglines. *KH*

Westfield – Scotland

The Westfield site was reputedly the largest opencast coal site ever worked in the United Kingdom. Located just west of the village of Kinglassie in Fife, Scotland, the site yielded some 26 million tons of coal between 1961 and 1986. It also claimed title to Europe's deepest open pit coal mine, reaching 850 feet below ground level at its deepest point. The site was worked by Costain Mining Ltd. in four phases under separate contracts from the OE.

The geology of Westfield was unique. The site boasted a low overburden-to-coal ratio with a total coal thickness of approximately 126 feet! But the many separate coal seams were interbanded with irregular partings of sandstone, shale and fireclay, together forming a self-contained basin, positioned well above the normal succession of coal found in the Fife area. The existence of this geological anomaly had been known for 150 years but since there was no continuous layer of coal workable by underground methods, its recovery had to wait until the development of large-scale earthmoving equipment.

Before production could start, the OE undertook preparation work as early as 1956, with contracts awarded to build a coal washery, roads and offices. A year later Costain-Blankevoort International Dredging was awarded a contract to remove the 4.3 million cubic yards of peat, silt and sand that lay over the site. A cutter suction dredger was used to pump the material as a slurry into permanent lagoons, which themselves were formed by placing 3.8 million cubic yards of compacted material in berms. The dredger carried a 7½-foot diameter rotating cutter head mounted on a 71-foot arm and weighed 550 tons in operation. It broke up the material so that it could be pumped in suspension through trailing pipes.

The dredging method of removing overburden proved most efficient, and the daunting project was completed without difficulty ready for mining to begin.

The first mining contract was for the recovery of seven million tons of coal and removal of 34 million cubic yards of overburden. A large fleet of shovels and dump trucks was selected, and the site was worked as one large open pit. This first phase involved creating a large void in which to dump spoil from subsequent phases. The material from the pit was dumped into hoppers, then crushed and transported two miles by conveyor for delivery via a mobile spreader to form a permanent landscaped area. The crusher reduced material to minus ten inches and was able to handle a throughput of 2,200 tons per hour. The 48-inch wide conveyor ran at a speed of 810 feet per minute, and consisted of several sections complete with transfer conveyors and drive stations. On completion, the fill had raised the original level of the barely cultivatable valley by an average of 110 feet. After surface soil placement, the area was developed into productive agricultural land by the Scottish Department of Agriculture.

Costain was awarded further contracts for coal extraction that ran until the main pit was completed in 1986. Excavation was initiated by nine electric shovels comprising Bucyrus-Erie 150-B and Ruston-Bucyrus 150-RBs with six-yard dippers and Ruston-Bucyrus 54-RBs with 2½-yard dippers. As further contracts were awarded over the years, the smaller electric shovels, intended for coal loading, were replaced with three diesel-powered Ruston-Bucyrus 71-RBs fitted with 4½-yard coal dippers. By 1976, two six-yard Lima 2400s, four more seven-yard Ruston-Bucyrus 150-RBs and a 12-yard 195-B shovel were added. Some of the electric 150-RB shovels were occasionally employed as coal-loaders in the thicker seams, likely the largest coal-loading shovels ever employed on a British opencast coal site.

To keep all the shovels busy, Costain employed an extensive truck fleet at Westfield, with more than 50 haul trucks running at any one time. In the early 1960s, Euclid R-22 trucks were the primary haulage vehicles. As bigger capacity haulers became available, the 22-ton units were replaced with 35–45-ton Euclid R-35 and R-45 models. The R-35s were fitted with coal bodies holding 32 cubic yards, and the 45-ton Euclid R-45s carried 36-yard bodies. By the end of the 1960s, most of the haulers at Westfield had been replaced with Euclid or Terex R45, 45-ton models built in Scotland.

During the 1970s, Terex R-50 trucks appeared as the haul truck fleet was continually modernized. However, truck size remained at 50 tons with the introduction of Terex 33-07, Aveling-Barford 50-ton Centaur and Caterpillar 50-ton 773 models in later years.

As with most large sites, Westfield boasted an extensive fleet of all types of support equipment. These included Gardner Denver blast hole drills, Michigan 280 wheel dozers, Komatsu D155A and D355A crawler dozers, O&K and Aveling-Barford graders and various smaller dozers. When Costain finished its contracts at Westfield, some 172 million cubic yards of overburden had been removed and 26 million tons of coal recovered.

Figure 7.135 The cutter suction dredger used by Costain-Blankevoort at Westfield to remove 4.3 million cubic yards of peat. The cutter head was 7½ feet in diameter and the machine weighed 550 tons. *Costain Mining*

Figure 7.136 This Muller spreader was used in the first phase to deposit some 32 million cubic yards of material to create new permanent farmland. *Costain Mining*

Figure 7.137 One of the initial Bucyrus-Erie six-yard electric shovels used for many years at Westfield.

Figure 7.138 An extensive fleet of Euclid R-22 trucks supported the excavator fleet, which included up to nine of these 150-B and 150-RB electric shovels.

Figure 7.139 Pit activity shows a Euclid R-45 hauler overlooking two 150-RB electric shovels and a multitude of machines in the deep pit beyond. *Costain Mining*

Figure 7.140 A Lima 2400 diesel shovel works on the upper benches of the massive Westfield pit. *Costain Mining*

Figure 7.141 An overall view shows the complicated geology of the Westfield open pit, which reached a depth of 850 feet below ground level. *KH*

BRITISH OPENCAST COAL

Epilogue

As the title indicates, this book focuses on opencast coal-mining from its origin in the UK up to 1985. Since that year, opencast production increased to a record level of more than 19 million tons in 1990, remained at a high level throughout the 1990s, then gradually tapered off to 8.6 million tons in 2013. During that time, a number of significant events took place, so a short Epilogue is included here to bring the story up-to-date.

As outlined in Chapter 6, hydraulic excavators gradually took over excavation duties from former cable machines. By the end of 1991, a British Coal Opencast report shows 117 hydraulic excavators of operating weight greater than 100 tons working on opencast coal sites in the UK. The most popular of these were: O&K RH-120C at 47 units; Demag H185 at 14 units; and O&K RH 90C at 11 units. Other various models represented were: Demag 13 units, Liebherr 13 units, O&K 12 units, Hitachi five units and Case Poclain two units. In addition, the report shows 332 hydraulic excavators under 100-ton weight class, mostly used for coal loading and cleaning. The top seven manufacturers of these, in order of popularity were: O&K, Caterpillar, Komatsu, Liebherr, Akerman, Hitachi and Hy-Mac.

The most notable change in on-site haulage was an average increase in dump truck size. The 50-ton Terex R-50, standard for many years, became a little small for the 12–15-yard shovels. Caterpillar's 85–100-ton 777 haulers were imported in increasing numbers as they were a perfect match for the most popular O&K RH-120C excavator. Even larger sizes were needed for the 30-yard class RH-200 excavators operated by Crouch and Budge, and by 1991 around 40 Caterpillar 789s of 190 tons capacity were operating in the UK. Articulated dump trucks in the 30–40-ton range also found favour for on-site haulage of coal because they manage adverse haul roads better than alternative road-going lorries.

In March 1988, coal production at St Aidans site in Yorkshire came to an untimely halt when an uncharted geological fault caused the River Aire to breach its banks and fill the open pit to a depth of 230 feet. Over the next few years, BCO worked with closely with the British Waterways Board and several other affected statutory bodies to ensure the impact of the failure outside the site was kept to a minimum. Several alternative schemes were considered, all involving the diversion of the river to a new course. By 1993, five years after the inundation, a plan was approved and British Coal Opencast awarded Alfred McAlpine Construction a contract to divert the river and adjacent canal into one single channel 2.2 miles long. When this two-year, £16 million contract was completed, the next phase of mine recovery could start. Now with new owners (RJB Mining took over British Coal Opencast in 1994), the site's massive void could be pumped out, a task taking a further two years.

In 1997 RJB awarded the St Aidans Remainder contract to Miller Mining and coaling restarted in February the following year. Miller continued to employ the Rapier W2000 dragline but not the old Bucyrus-Erie 1150-B. This machine, known as 'Oddball' was preserved as a permanent tourist attraction at the site. The remaining coal extraction was completed by 2002, and the following year the site was transformed into a country park for public recreation. Before the flood, the site yielded 3.7 million tons of coal, and the Remainder contract yielded a further 2.8 million tons.

In 1987, the Opencast Executive of the National Coal Board (NCB) was renamed British Coal Opencast (BCO). In late 1994 the UK coal industry was privatized, and Budge – now known as RJB Mining (UK) Ltd. since Richard Budge took over as managing director – acquired the entire assets of the former National Coal Board underground and surface mining operations in England. At the same time Celtic Energy Ltd. took over NCB assets in Wales, and in a similar arrangement Scottish Coal acquired NCB assets in Scotland. In June 2001, RJB Mining became UK Coal.

Following financial difficulties, UK Coal was restructured and became Coalfield Resources plc in 2012. From July 2013 to November 2014 the company was known as UK Coal Surface Mines, and the current owner, from November 2014, is UKCSMR Ltd., which was incorporated on 22 October 2014 and is based at Harworth, Doncaster. In Scotland, Scottish Coal was placed in liquidation in April 2013 and all sites temporarily shut down. Since that date, sites have been managed by Hargreaves Services plc, which gradually reopened them and took ownership of the former assets of Scottish Coal.

Figure EP1 The Bucyrus-Erie 1150-B named 'Odd Ball' that last worked at St Aidans site is on permanent display here, preserved by local volunteer group, the St Aidans Trust, who took over the machine in 1999. *KH*

Figure EP2 After her time at Radar North, this Marion 7800 dragline (serial no. 9997) worked at three more sites – Radcliffe, Togston and Chester House – and was finally scrapped in 1994. The other Marion 7800 (serial no. 9998) was sold in 1969 to Canada where it opened up the new Highvale Mine west of Edmonton, Alberta. After a productive life of 51 years, the old 7800 workhorse was finally parked in 2007, as seen here, and scrapped in 2013. *KH*

Figure EP3 The Rapier W1800 dragline at Maesgwyn worked there from 1961 to 1988, when it was sold to Beltrami Enterprises and shipped to Pennsylvania, USA. The unassembled parts lay there at an opencast site now operated by UK company Coal Contractors (1991) Inc., until they were finally scrapped in 2003. *KH*

240 EPILOGUE BRITISH OPENCAST COAL

Figure EP4 This 20-yard Marion 7500 dragline was one of the stripping machines at Fairclough-Parkinson's Nant Helen site in South Wales, starting work here in 1989. The machine was actually purchased new in 1979 to work in the opencast ironstone industry at Scunthorpe. But because of the British Steel Corporation's decision soon afterwards to discontinue home-produced iron ore, the machine only worked there briefly. It was moved to Nant Helen and, on completion of this site, sold to Blaschak Coal Corporation and moved to Pennsylvania.

Figure EP5 When the 1550-W 'Big Geordie' moved to Butterwell in 1977, its owner, Derek Crouch, leased it to Taylor Woodrow until 1991, for the duration of that contract. No doubt Crouch hoped for further work for its biggest excavator at the nearby Stobswood site. But the OE had already ordered and planned to use its own new P&H/Page 757 'Ace of Spades' instead. Big Geordie worked no more and was parked with boom lowered. The author took this photo in 2003, a few months before the 3200-ton machine was scrapped.

Figure EP6 Miller Mining's Kirk site in Derbyshire just after completion in 1997. Most of the area in the picture has been opencast mined and awaits final seeding and cultivation. Three Ruston-Bucyrus 195-B shovels are ready to move to new destinations. *KH*

Figure EP7 In 1992, the OE's newest and largest walking dragline was commissioned. Amid fireworks and major press coverage, 'Ace of Spades' – the largest dragline in Europe – was launched at Stobswood site near Morepeth, Northumberland. The 4,400-ton P&H 757 giant digger swung a 65-yard bucket on a 310-foot boom. This picture was taken in 2003 during the last few days of its operation at Stobswood. *KH*

242 EPILOGUE BRITISH OPENCAST COAL

Figure EP8 The 'Ace of Spades' Page 757 dragline that worked at Stobswood opencast site is now re-named the Liberty Belle. It will remove the overburden sitting on top of the Kemper County's supply of lignite, and began operating in mid-November 2013. *Photo taken: October 2013. Copyright Mississippi Power. All rights reserved.*

Figure EP9 An example of how opencast coal-mining sites are restored to natural countryside. Hedgerows in the foreground and fields in the background were planted following mining in 1989 at the former Ryefield site in Derbyshire. *KH*

EPILOGUE 243

Figure EP10 Another example of restored land is this former Street Lane site near Denby, Derbyshire.

Figure EP11 Taken at the still-active Lodge House opencast site at Smalley, Derbyshire, this 2014 view shows fields re-established in the first phase completed just four years earlier. *KH*

Author's view

Over the past two decades the subject of coal-mining has become a highly political issue due to the global warming debate. This delicate subject has caused governments in certain countries, and especially the UK, to introduce legislation precipitating the rapid decrease and ultimate phase-out of coal.

Too often we encounter studies from some group or institution, learned or otherwise, on the topic of climate change or 'global warming'. These studies often report widely differing results. Views range from taking immediate steps to save the planet from rapidly rising temperatures and oceans, to statistics that the earth's temperature has actually been cooling, or at the very least remained constant for 15 years or so, despite an increase of carbon emissions during this period.

With no chance to participate in these various studies, most of us can only digest the information and form our own opinion of what may or may not be happening to our climate. I don't necessarily support burning coal to generate our electrical energy, but I know that coal is currently the cheapest and most reliable form of energy generation. I also know that the climate is changing, but it always has since the dawn of time. So, until undeniable evidence proves that carbon emissions affect our climate and a reliable, economical and safe alternative source of energy is found, I believe that governments should not legislate expensive alternatives to coal. After all, carbon is not a pollutant.

While UK-produced coal has decreased in recent years, it is ironic to report that imported coal has increased, meaning that total coal consumption has only marginally decreased. In 2014, UK government statistics show net imports (after deduction of exports) were 44 million tons, while home-produced coal totalled only 12.7 million tons (8.6 million opencast, 4.1 million underground). In other words, home-produced coal represented only 22 per cent of total UK coal consumption.